FUNDAMENTALS
of
ENVIRONMENTAL
SCIENCE and
TECHNOLOGY

Edited by
Porter-C. Knowles, P.E., P.G.

Government Institutes, Inc.
Rockville, Maryland

Government Institutes, Inc., Rockville, Maryland 20850.

Printed in the United States of America

Library of Congress Cataloging-in-Publication Data

Fundamentals of environmental science and
    technology / edited by Porter-C Knowles (Dames
    & Moore).
            p.   cm.
    Includes bibliographical references and index.
    ISBN 0-86587-302-X : $24.95
    1. Environmental engineering. 2. Water quality. 3. Air
quality. 4. Soils--Quality. I. Knowles, Porter-C. II. Dames
& Moore.
TD145.F86 1992
628--dc20   92-71886
                                                                 CIP

Printed on acid free paper

CONTENTS

• • •

ABOUT THE AUTHORS

• • •

Linda L. Black-Covilli

Ms. Black-Covilli holds a B.S. in chemistry and spent her early career as a research chemist with Monsanto Company in St. Louis, in the Agricultural Research Division. She synthesized compounds which were the precursors of Roundup™ herbicide. In 1977, she took a position as environmental chemist with ACF Industries and was later promoted to corporate manager of chemical safety and environment. She worked in an environmental/safety/industrial hygiene capacity for thirteen years. Since May of 1990, Ms. Black-Covilli has managed the Kansas City office of Dames & Moore. She has expertise in the performance of environmental compliance assessments, having experience with the chemical manufacturing industry as well as other manufacturing industries.

Thomas M. Covilli, CIH

Mr. Covilli holds a B.S. degree in zoology and is a registered certified industrial hygienist. Mr. Covilli began his career as environmental chemist for ACF Industries in St. Louis, Missouri, where he developed company-specific material safety data sheets and safety-related procedures. After five years at ACF and promotion to senior environmental chemist, Mr. Covilli took a position with Petrolite Corporation for two years, after which he developed his own consulting firm, Allstates Environmental Services. At Allstates, Mr. Covilli provided industrial hygiene, Health and Safety training, and environmental remediation services. In June of 1990, he joined Dames & Moore as an associate and became the regional health and safety manager for the mid-continental United States. As a consultant with Dames & Moore, Mr. Covilli continues to provide industrial hygiene and training services in addition to project management in large environmental remedial projects. He has also participated in human health risk assessments for two large remedial efforts.

Robert G. Cooper

Robert G. Cooper is a mechanical engineer specializing in assessment, construction management, and project management of contaminated sites. He is currently an Associate in Dames and Moore's Southern Region, managing the remediation practice in Florida and the Caribbean. He received his B.S. in applied science and engineering from the University of Toronto Canada in 1973. Mr. Cooper has worked in the United States, Canada, the Caribbean, South America, and the South Pacific.

Kaushik Deb

Mr. Kaushik Deb is an environmental engineer in Dames & Moore's Chicago office. He joined Dames & Moore in 1989. Mr. Deb specializes in air pollution control (APC) technology evaluations, equipment selection, and air permit applications for both criteria and toxic air pollutants for a wide variety of industrial facilities. He has wide-ranging experience in source sampling, top-down BACT/LAER (Best Available Control Technology/ Lowest Achievable Emission Rate) analyses, APC system design, air emissions inventory development, and air dispersion modeling. As Principal Investigator on all of the toxic air pollutant projects performed by Dames & Moore for 13 Wisconsin pulp and paper companies, and on two projects for a domestic and foreign automaker in Wisconsin and Kentucky, Mr. Deb has considerable experience in air emissions estimating, permitting, and compliance in the Midwest.

Perry W. Fisher, Ph.D.

Dr. Perry W. Fisher is a Principal and Certified Consulting Meteorologist at Dames & Moore's Chicago office. He joined Dames & Moore in 1973. Dr. Fisher specializes in air quality regulations and permitting as well as toxic air pollutant assessments. He has particularly extensive experience with the pulp and paper and automotive industries. During his career with Dames & Moore, he has directed more than 50 projects for pulp and paper companies, 75 projects for automotive companies related to air quality, and more than 30 prevention of significant deterioration (PSD) permit applications for a variety of industrial facilities, primarily in the Midwest. He has served in a variety of capacities at the local and national level of the Air and

Waste Management Association (AWMA), the most recently as chairman of the Sections Council. Since the passage of the 1990 Clean Air Act Amendments (CAAA), he has taken the lead for Dames & Moore in speaking at Dames & Moore seminars and other presentations across the nation for a wide variety of industries and in Japan summarizing these amendments and recommending how industries can be pro-active in responding to them. Dr. Fisher has published over twenty-five publications relating to air quality.

Neil J. Jost, Jr., P.E.

Neil Jost is an environmental engineer with twenty years experience in monitoring, treatability, design, and project management. He has worked with a broad range of wastewater and ground water problems for industry as an environmental manager and consultant. Two recent projects were review of the use of denitrification for removal of raffinate pit nitrates at the Department of Energy Weldon Spring NPL site and development of a glycol spill cleanup plan to remedy soil and ground water contamination at a food processing plant. He is an associate and office manager for Dames & Moore's St. Louis office and a registered professional engineer in Missouri. He attended St. Louis University (B.S.) and Washington University (MS) in Saint Louis.

Porter-C. Knowles, P.E., P.G.

Porter-C. Knowles, P.E., P.G. is a principal in Dames & Moore's Kansas City office. Mr. Knowles has over 25 years of experience involving a wide variety of multi-discipline projects, including ground water and hazardous waste contamination investigations. Being both a registered professional geologist and professional engineer, Mr. Knowles has a particular appreciation of multi-discipline projects and the need for individuals to gain a basic understanding of many areas of science and technology to solve today's regulatory problems. Mr. Knowles attended Yale University and the Colorado School of Mines (M.S.). He is a member of numerous professional societies and regularly speaks and publishes on technical topics. Mr. Knowles joined Dames & Moore in 1968 and was a member of Dames & Moore's board of directors from 1987 to 1991.

James W. Little

James W. Little is a senior air quality consultant and associate with Dames & Moore. His responsibilities include air quality permitting, air quality modeling, toxic air pollutant risk assessments, environmental audits, and project management. He has 23 years of experience in air quality and atmospheric sciences. His educational background includes an M.S. in air and industrial hygiene from the University of North Carolina.

Andrew P. Schechter, P.E.

Mr. Andrew P. Schechter is registered as a professional engineer in Florida, New York, New Jersey, and Pennsylvania. He is a principal in Dames & Moore and manages their waste management group in the Southern Region. He received his B.S. in civil engineering in 1972 from the Polytechnic Institute of Brooklyn and his master's degree in civil engineering from the Polytechnic Institute of New York in 1974. Mr. Schechter has performed numerous contamination assessments and remedial action programs throughout the eastern United States.

Robert J. Taylor

Robert J. Taylor is a petroleum engineer, with eight years of petroleum industry experience and four years as an environmental consultant with Dames & Moore. Mr. Taylor is currently a project manager with Dames & Moore, responsible for soil and ground water assessment and remedial system design and installation. He has worked in the Caribbean, Central America, and Korea. He received his marine science degree from Florida Institute of Technology in 1979 and a bachelor's degree in science and technology from Nicholls State University in 1987.

Gary F. Vajda, P.E.

Gary F. Vajda, P.E., is the managing principal of the St. Louis, Missouri office of Dames & Moore. Mr. Vajda directs environmental audits; regulatory consultation and strategies; site investigations and remediations (RI/FS); and pollution prevention evaluations for a number of corporate and facility clients in the midwest and elsewhere in the United States. Mr. Vajda has more than 18 years experience in environmental engineering. Previous responsibilities have included process and engineering design, project management, and regulatory liaison in the engineering/construction and manufacturing industries for projects dealing with air, water, and hazardous waste. He has worked on projects in the chemical, petrochemical, iron and steel, and general manufacturing industries, and has worked in the United States, Great Britain, and Poland. Mr. Vajda received his M.S. in environmental engineering from Illinois Institute of Technology, Chicago. He is a member of the Water Pollution Control Federation and the Air and Waste Management Association. Mr. Vajda is a registered professional engineer in Missouri, Illinois, Indiana, Ohio, Kentucky, Michigan, and Louisiana.

David C. Van Dyke

David C. Van Dyke is an associate with Dames & Moore in its Kansas City office. Mr. Van Dyke joined Dames & Moore in 1990 after working since 1973 in heavy manufacturing and foundry industries. He has managed projects under RCRA, CERCLA, CAA, and CWA. A current project is the development of a totally enclosed treatment system to be applied to the basic steel manufacturing industry. Mr. Van Dyke earned his B.A. from the University of Wisconsin in 1967, flew as a foreward air controller (FAC) in Vietnam, and earned his M.B.A. in 1989 from Keller Graduate School of Management.

DAMES & MOORE

Dames & Moore is a worldwide professional firm providing consultation in planning, engineering, the earth and environmental sciences, waste management, design, construction management, and regulatory assistance. Founded in 1938 by Trent Dames and William Moore as a partnership, Dames & Moore now is a publically owned company listed on the New York Stock Exchange. Dames & Moore has over 100 offices worldwide and a staff of over 3500 personnel. Principal clients include businesses and industries in the private sector, public and private utilities, financial institutions, developers, architect-engineers/constructors, and all levels of government. The firm is ranked 16th out of the top 500 Design firms by *Engineering News Record* (ENR) and 10th of the top 75 environmental consulting and engineering firms by *Environment Today* (July 1992).

PREFACE

• • •

Environmental problems and issues today involve a number of interrelated scientific and technical disciplines. Therefore, to solve the problems and understand the issues, managers need a broad base of information from which to draw.

That has been the challenge of preparing this first edition of *Fundamentals of Environmental Science and Technology*: to effectively bridge both the science and technology involved; to offer coverage of the topics that is broad but sufficiently detailed to be of value, and, in doing so, not to needlessly envelop that detail in technical jargon. In short, our goal has been to focus, in plain language, on the big picture while serving as a launching point for further knowledge and insight. Consequently, this book will hopefully serve just as varied an audience: students, lay-personnel, technical experts, attorneys, regulators, and policy makers.

In forming this bridge between basic science and technology, this book explains how we apply that science to get basic data and information on the existing conditions of air, water, and soil. It then discusses the technologies currently used to address environmental problems of solid and hazardous waste, air emissions, and water quality.

Although we have tried in this book to cover an incredible amount of information using common language rather than technical jargon, there is, nevertheless, a certain amount required for basic communication in these fields. Therefore, we have used acronyms with which we think most people will already be familiar, such as "BAT" for "Best Available Technology", "RCRA" for "Resource Conservation and Recovery Act", and similar terms.

As the following chapters are studied, readers should develop a new understanding of the various basic scientific and technical fields, and see more clearly how all these different sciences and technologies relate to each other and to compliance with environmental regulations. Readers should consult the references listed at the end of various chapters for additional information on a particular topic.

Given the fast-moving developments related to technology, treatment concepts, and new equipment, a new edition will be inevitable. In the meantime, please write us with your thoughts about how well we have met the goals of this first edition.

Porter-C. Knowles

Chapter 1

• • •

GEOLOGY AND
GROUND WATER HYDROLOGY

Porter-C. Knowles, P.E., P.G.
Dames & Moore

OVERVIEW

The focus of hazardous and solid waste legislation has been and is an effort to clean up our waters—both on the surface and below ground. Ground water, in particular, presents the biggest problems as well as the biggest benefits. To understand where ground water is, under what subsurface conditions it exists, and what the level of water contamination is are questions scientists and engineers help to resolve. The solutions become expensive and technically complex. Surface waters also pose problems of assessment and control. Some efforts such as Lake Erie and the Everglades in Florida are massive and mind boggling in scope, level of effort, and cost. Many times, the problem itself is extremely difficult to even understand and comprehend.

July 1992 press releases from the Office of Management and Budget (OMB) state that Superfund (CERCLA) cleanups are costing over 30% more than estimated by the EPA. The staggering costs of cleanup, particularly in the subsurface, have raised concern among many parties who, in economically tough times want to feel their tax dollars are wisely spent. Are cleanup levels realistic? Should cleanup be based upon more of a risk assessment system than absolute numbers? Superfund PRP (potentially responsible parties) groups may spend inordinate amounts of funds and energy trying

1

to locate *other* potential PRP's rather than in cleaning up the problem. Should the government offer more "carrot" than "stick" to those willing to clean up problems promptly?

As a basis for understanding the problems of pollution and the technology used to address contamination problems, it is appropriate to start with the subsurface—the sciences of geology and ground water hydrology. The following sections and chapters will serve to raise your understanding of basic issues and, in turn, help the reader with environmental policy and perspective.

GEOLOGY

The earth beneath us is the fascinating story of geology. As we go from mountains to the coastline, from desert to the wetlands, we see around us the surface of the earth being shaped in many different forms by a wide variety of external and sometimes internal forces. It is the weathering of rocks that provides us with an immense diversity of landscape. It is also the rocks which provide the medium through which water flows—a water which we call ground water, providing ninety percent of man's water supply. To understand the presence and the movement of ground waters, we need first to have a basic understanding of geology, which is the study of the material through which ground water moves.

To start with, there are three basic types of rocks and a large number of soils resulting from the weathering of these rocks.

The first rock type is *igneous*—rock material which at one time was in a molten or liquid stage. Typical igneous rocks are the many types of volcanic material which all of us have seen in pictures or in person—volcanic eruptions spewing from a volcano, flowing down a hillside, and later solidifying into rock. A less obvious type of igneous rock occurs when molten material solidifies beneath the surface of the earth rather than on top of it. This type of rock is found beneath the volcano and beneath the earth's surface where a continuing melting and solidifying occurs at depth—caused by the heating and cooling of the earth itself from its core. A good example of this type of rock is granite found in the core of many mountain areas. Granite is the light speckled rock seen in many mountain road cuts. We are able to tell the origin and history of rocks by physical examination of a thin section of rock with either a ten power hand lens or a microscope. Dependent upon the chemical composition of the molten lava or magma and the cooling time, igneous rocks are categorized by the coarse or fine texture of crystals or grains. The chemical composition of the materials also typify the source of the rock.

The second type of rock is *sedimentary*. Sedimentary rocks have one thing in common: they are composed of small units, ranging in size from molecules up through dust particles to pebbles and large boulders brought together and deposited on the surface of the earth's crust (above or below water). Some components were transported by water, and others by wind or glaciers or gravity. All of the mineral matter composing these rocks was once part of other rocks—igneous, previously existing sedimentary rocks, and metamorphic (to be discussed next). It should also be pointed out that some of these materials may have passed from solution in water prior to their becoming part of rock material. The consolidation and/or cementing of these particles is sometimes described by the term *lithification*.

Most but not all sedimentary rocks are stratified, having been put down in layers or beds. However, not all stratified rocks are sedimentary (volcanic lavas are often stratified). The physical processes causing the initial weathering may also have impact on how well the particles are cemented together. It is a cementation of the materials which is the difference between rock and soil. It follows that some soils (soil particles) become cemented together after their deposition due to chemicals left by ground waters in many situations. The particles which form sedimentary rocks and which, when uncemented, are called sediments range in size from clay through silt through sand (fine, medium and coarse) through pebbles, cobbles and boulders. As you can imagine, the texture of sedimentary rocks is dependent upon the various mixtures of the different sizes of particles. Why is this important? A well sorted sandy material with particles very nearly the same size will have a much larger space between particles (called porosity) and will also allow better connection of those pores, which when we measure that connection is called *permeability*—the ability of rock to transmit liquids or air.

Let us examine for a minute, a piece of sandstone. Sandstone is one of the most frequently observed sedimentary rocks, not because it is more abundant than other types, but because it has a tendency to appear in prominent cliffs, forming the walls of canyons, ledges and shoreline. Generally, you can determine the bedding planes which separate the layers of sand as part of the deposition process. These planes, or parallel surfaces, are weaker than the rest of the rock and are usually etched more deeply by weathering and erosion. Using a magnifying glass, the sand grains that make up the bulk of the rock may range in size and show evidence of cementation which in many cases is the rusty or yellowish color of iron oxide with some calcium carbonate. Calcium carbonate is a typical cementation, resulting from sandstone deposited in fresh and salt waters and/or later through movement of ground water through soils.

Another common sedimentary rock which is composed mainly of calcium carbonate (not just a cement between sand grains) is limestone. Carbonates dissolved in fresh or salt water may be precipitated in solid form in many ways. In the most common processes, aquatic animals and plants secrete the calcareous material to construct their shells, bones, teeth etc. This biochemical precipitation may sometimes result in limestone beds consisting almost wholly of fragments of sea shells more or less broken and worn by waves and currents. Other limestones are obviously ancient reefs built by corals and other associated animals and plants. In chemically analyzing the rock, many limestones contain a significant amount of calcium-magnesium carbonate rather than just calcium carbonate. When magnesium carbonate is the primary constituent, the rock is called dolomite.

The third and final category of rocks is *metamorphic*. The term metamorphic means simply "changed in form." The name concentrates attention on the processes by which the rock evolves. All of the rocks in this class were once either igneous or sedimentary but have been changed by pressure, heat or chemical action of liquids or gases so that their original nature has been significantly altered. Pressure is caused by burial of rock types beneath the surface, and heat is applied either as a function of depth or of close proximity to molten rock from magma beneath the surface or igneous intrusions into the upper rock layers. Depending upon the chemistry of the rocks involved, new crystals and mineral compositions occur which are only possible under temperature or pressure. For example, rocks buried beneath the gulf of Mexico together with temperature eight miles below the earth's crust represent enough forces to start the process of metamorphism. Some typical metamorphic rocks and the parent material which you will easily recognize would be limestone turning into marble, or shale (silt sized sedimentary material) changing into slate. Sandstone, for example, in many cases is turned into quartzite due to a silica rich water environment, pressure and temperature.

Metamorphic rocks are important because the processes of metamorphism sometimes concentrate high grade metal bearing ores. Magma at a

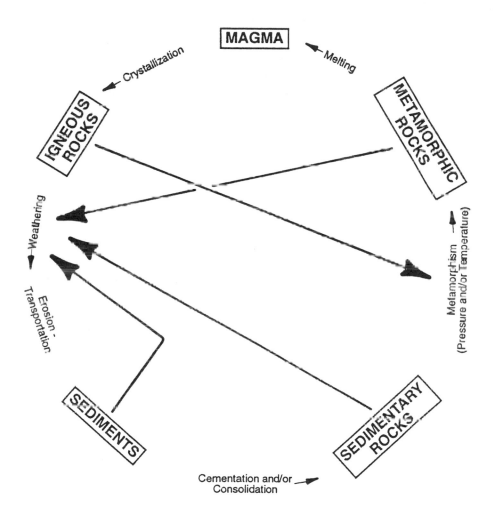

Courtesy: Dames & Moore

FIGURE 1
The Metamorphic Cycle

depth adjacent to iron rich sandstones may under the proper conditions result in concentrations of iron sulfides and other metal sulfides. The most common metal ores are sulfides.

A brief comment about the concept in geology called "the metamorphic or rock cycle." A given particle of material may travel completely around the cycle starting as a fluid, solidifying as an igneous rock, being broken down into a sedimentary rock which then may undergo metamorphic processes later to return as molten magma or other sediments. This cycle is a continuous operation over the earth's surface with an incredible number of variations. It is the geologist who, through study of rock types, evaluates where the rock is in the cycle, and how it reached that point. With an appreciation for the erosional processes of rocks and depositional characteristics of rocks, a geologist can understand why topography is the way it is and can map rock types by topographic surficial features. Why is this important? In our context today, the hydrology relating to ground water is controlled by the physical environment beneath the earth's surface. The extent of pores between sedimentary particles, the fractures caused by rock stress or the "solutioning" of material provide not only space for water but a mechanism for water to move beneath the surface as part of the "hydrologic cycle." This new cycle leads us into discussion of the science of ground water.

GROUND WATER HYDROLOGY

The *"hydrologic cycle"* can be just as varied as the previously discussed metamorphic and/or rock cycle. Basically, the hydrologic cycle provides a simply illustrated pathway starting in the atmosphere. Water is then deposited on land and water by either rain or snow onto the surface. On the ground, some water flows in surface runoff into lakes, streams, rivers and the ocean, and as other waters move into the earth beneath us as ground water. Ground water movement for the most part reflects topography with water continually moving down gradient to a point of equilibrium which in many cases is the ocean. Water returns to the atmosphere through evaporation of surface waters and through transpiration from plants which bring ground waters to the surface by their root systems.

In looking at a simple picture of a runoff cycle, water infiltrates through surficial soils to a "water table." The water table is the surface below which soils are saturated. Usually above the water table there is an intermediate or

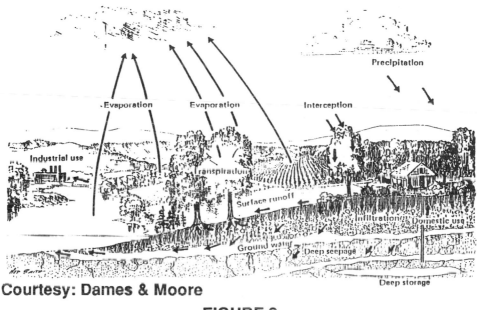

Courtesy: Dames & Moore

FIGURE 2
The Hydrologic Cycle

"vados zone" which is not completely saturated. Once into the water table, subsurface formations may be designated as *aquifers*. Some saturated formations are so impermeable they are called *"confining units"*. For confining units, the formation may be saturated but represents a change in the ability of the soil to transmit water. For example, clays and silts will provide a barrier to ground water movement and are usually considered confining units. In contrast, more permeable material such as sands, sandstone, gravels etc. comprise good aquifer units. Aquifers store and transmit water in significant amounts to adjacent beds or bedrock or to the surface as springs. For ground water supply, wells are installed in aquifer units to retrieve good quality water in sufficient volume to meet demand.

Before we discuss ground water hydraulics, it may be helpful to evaluate various formations and rock types as to their *permeability* and *porosity*—their ability to store (porosity) and to transmit ground waters (permeability)—which may be used for water supply. A water supply may be of high quality for drinking or of lesser quality for industrial use or for cooling. The following table illustrates some water-bearing properties of common rocks showing their permeability and porosity. Permeability or hydraulic conductivity (K) depends on properties of the fluid as well as the characteristics of the medium.

Average Values of K

Soil Class	K, cm/sec
Gravel	1 to 10^2
Clean sands (good aquifers)	10^{-3} to 1
Clayey sands, fine sands (poor aquifers)	10^{-6} to 10^{-3}
Silts and clays (confining units or cap material)	10^{-2} to 10^{-7}

Note the "orders of magnitude" between the various types of soil. For example a gravel aquifer may be 100 to 1000 times more permeable than a clean sand aquifer.

Again, porosity is the pore space available for water storage between rock or material particles, whereas permeability relates to the connection of the pores or the ability of water to flow through rock material. In general, sands, sandstones and gravels will have much higher permeability than silts, clays or crystalline rock.

Very much like the gradient on the surface of the earth, the steeper the slope, the faster the movement of ground water through rock or soil material. *Gradient* as well as permeability/porosity are important parameters to know in evaluating ground water movement. Although gravely riverine materials adjacent to the river may have very high permeabilities and porosities, the gradient or slope may be very flat. This combination may result in very slow movement of ground water adjacent to those rivers and streams. On the other hand, ground water as well as surface water in mountainous terrain with high gradients or slope will move much more quickly to streams or the water table. Water moving through soils and/or rocks above the water table move at a gradient of one (1). This gradient is gravity movement of liquid water through saturated sediment. In moving water through the subsurface, you can understand that water may run into layers of less permeable materials such as silts or clays which beneath the surface will "pond" the water until the water flows horizontally and until it reaches more permeable material and can proceed downward to a lower water table elevation or equilibrium. This particular situation, common in many localities, is called a "perched" water table. The perched water table may be of significant thickness or may be so thin as to be marginal with respect to water withdrawal.

Although this discussion has primarily emphasized sediments and sedimentary rock materials, water clearly can move through the subsurface through faults, zones of weakness in the rock or material, as well as fracture patterns in the geologic strata. Water can also move from the water table via a spring to the surface and then back into the ground water again.

We have mentioned gravity and slope or gradient as a factor in moving ground waters. Another factor impacting movement and related to contamination of ground waters is density. For example, saline waters or salt water/sea water, being more dense than fresh water, will underlie fresh water when both are present under equilibrium conditions. Islands in the ocean normally will have a fresh water lens resulting from rainfall which basically "floats" above the saltwater surrounding the island environment.

When we look at contamination in ground waters, an important factor is whether the chemical constituents are (1) soluble or insoluble in water; or (2) are more dense than the water itself or are significantly lighter. The light fractions will be found at the upper surface of the water table—known as "floaters." The more dense contaminants will tend to move through the ground water to a confining unit or barrier (they may move more quickly vertically than horizontally) and are known as "sinkers."

An important aspect of the physical and chemical properties of water relates to using chemical evaluation of waters to determine the source or origin of the water. There are several means of evaluating water origin, some utilizing radioactive constituents to age date the water and providing a perspective on infiltration compared to other ground waters. One technique also utilizes the chemical evaluation of anions and cations plotted in a triangle to evaluate a "signature" or "fingerprint." If a question is raised as to whether waters from different aquifers are mixing together through zones of weakness or connection, the use of tri-linear diagrams or Stiff diagrams facilitates rapid comparison of analysis of waters from different depths to evaluate their origin.

As you can surmise, ground water movement in the subsurface is primarily controlled by the permeability of the soils/rock (ability of the rock to move water) and the gradient or slope of the water table or in the case of artesian waters the potentiometric surface. Ground water velocities can range from several feet per day to several feet per year. In the cases of very tight confining units, movement may be tenths of a foot per day to a tenth of a foot over ten thousand years. The point should be made, however, that clay soil liners although highly impermeable if placed properly, will still facilitate movement of water although it is at a very, very slow rate. It is only with synthetic liners of impermeable materials that no movement would occur across such a barrier. It is also understandable that chemical constituents involving molecules much smaller than water could move through somewhat impermeable soil or rock material much faster than water. The installation of natural clay liner material decades ago many times did not take this physical fact into account, and assumed that these liners would prevent all movement. They do not.

After discussing some general geology and hydrology, we are at a point to discuss groundwater contamination—the identification, the evaluation and the clean-up. Why was all of the preceding background information important? For us to understand contamination, we need to understand the physical environment where that contamination can be found.

In fairly simple terms, an evaluation of a contamination problem starts with evaluating the physical characteristics of the subsurface.

By use of some surface geophysical techniques which may be able to identify a contaminant "signature," the general extent of both horizontal and vertical of contamination may be identified. We can utilize that information to provide a more specific testing program. Some geophysical techniques also help to define the characteristics of subsurface materials such as clays and sands which may provide either barriers or pathways to groundwater flow. Examples of geophysical techniques used on the surface, which have the advantage of covering large areas at a low cost, are measurement of magnetics, conductivity, resistivity, sound waves and ground penetrating radar. Each one of these methods provides an indirect means, through anomalies, of evaluating subsurface physical conditions. However a word of caution: an important aspect of geophysics is the ability to "ground truth" with specific measurements of one part of the subsurface to correlate with measurements over a large area. It should also be mentioned that down-hole geophysics utilized in borings and wells provide a similar tool to evaluate subsurface rock, soil and water components without taking specific soil, rock or water samples. This sampling is at a much greater cost. As mentioned, geophysics may provide the framework for qualitative information on which to make some sound decisions. More specific actions may then be taken in investigating the subsurface.

Drilling into the subsurface, actually collecting samples of both soil/rock and water quality as well as then "testing" the aquifer or confining layer to measure specific permeability of a zone is normally required to some degree for contamination studies. To adequately characterize the subsurface aquifers and confining units, in addition to specific evaluation of a contamination plume, may require a range of costs. Understandably, the complexity of the subsurface conditions, both geologically and hydrologically, greatly impact overall costs of specifically identifying what contaminant is located where. Whenever drilling and testing is required, invariably costs are above five figures, and in some cases, RCRA and Superfund sites have ranged into several million dollars or more.

Once the subsurface hydrology and geology have been defined as well as a contaminant plume, state of the art computer modeling techniques are also available to predict longer term migration of various contaminant levels. This modeling may be important in evaluating risk assessment relating to human pathways and environmental pathways. Dependent upon the quantity and quality of information, it is possible today to utilize three-dimensional modeling techniques for these predictive efforts. Modeling, however, is also not cheap. Five figure modeling efforts are common, with the more complex models going to six figures.

In conclusion, it should be clear that to understand the earth beneath us is many times a challenging, complex and expensive effort. Nevertheless, modern technology involving a number of primary and secondary measuring techniques ranging from satellite imagery to surface geophysics to subsurface drilling and testing can provide us information with which to evaluate, utilize and preserve our environment. Whether the search is for a long term, potable water supply; a search for non-potable cooling water for industrial plants; or for preservation of water levels and flows in the environment impacting surface ecology: these problems can be addressed with an appropriate technical and economic level of solution.

For those who wish to look further into the fields of geology and/or hydrology, there are a number of excellent references on the market which should satisfy readers of varying technical backgrounds. An excellent source of elementary materials may be found from the U.S. Geological Survey with offices in Reston, Virginia; Denver, Colorado; and many other locations. Your local EPA Regional Headquarters also has available a number of primers to help understand these fields, and your local State Geological Survey is an excellent source. Many State Geological Surveys also have roadside handbooks which explain in detail geology and sometimes hydrology across a particular state.

ADDITIONAL REFERENCES:

Ground Water and Wells, Johnson Division, Fletcher C. Driscoll, 1986.

Ground Water Handbook, Government Institutes, Rockville, MD, 1992.

Drinking Water Treatment Technologies: Comparative Health Effects Assessment, Government Institutes, Rockville, MD, 1990.

Miller, Leonard A., (ed.), *NPDES Permit Handbook*, 2nd Edition, Government Institutes, Rockville, MD, 1992.

Chapter 2

• • •

BASIC ENVIRONMENTAL CHEMISTRY OF HAZARDOUS AND SOLID WASTES

Linda L. Black-Covilli
Dames & Moore

OVERVIEW

Through the application of science and technology, *Homo sapiens*, like no other species, has been able to change his environment drastically. Many of these changes have improved his condition on earth. However, some of these changes have also produced adverse effects for all of us living on the earth.

Chemistry affects just about every area of our daily lives. Not just an interaction between atoms and molecules, it is important to other sciences as well. Chemistry applied to the science of biology has brought about many new advances. It has helped us to understand better how living organisms function. Many of our environmental problems, such as cleaning up the air and water and treating our solid wastes, are also being solved by the use of chemical knowledge.

WHAT IS CHEMISTRY?

Most of us already have a good idea of what atoms are; they are the fundamental building blocks of matter. There are 105 different kinds of atoms known to science. Of these about 90 are naturally occurring. The rest have been produced artificially in research laboratories. Chemistry concerns the manner in which these 105 different types of atoms arrange themselves in the various forms of matter. It is also the study of transformations of one kind of matter into another.

Every object in the environment is either a chemical substance or mixture of chemical substances. The food you eat, the clothes you wear, the gasoline burned in your automobile and medicines taken when you are ill are all products of some kind of chemistry wrought either by the forces of nature or the hands of people. Your body is made up of chemicals, thousands of them, that are continuously reacting with one another in a manner that makes a busy chemical manufacturing plant look almost idle.

Chemistry in some way or another underlies practically all the other sciences. Consider a piece of copper wire. Copper is a good conductor of electric current; this particular phenomenon happens to lie within the realm of physics. Copper also reacts with nitric acid to produce copper nitrate; this is a chemical transformation. Both of these phenomena, one physical, the other chemical, are really quite closely related because both involve displacement of the outer electrons of copper atoms.

Chemistry has significance to zoology as well. Tigers have stripes while leopards have spots because of a chemical substance called deoxyribonucleic acid (DNA). It is found in the cells of virtually every living organism. Since DNA is responsible for transmitting inherited characteristics, tiger DNA, as expected, differs from leopard DNA.

As for the relationship of chemistry to psychology, the functioning of the brain is a predominantly chemical process. It is well known that when certain abnormal chemical changes take place in the brain or endocrine system, marked changes in behavior occur—in schizophrenia, for example, or the influence of drugs.

Chemistry has significance even to the humanities. The restoration of a priceless old master's painting frequently calls for a high order of chemical skill to repair the ravages of time. There is certainly a connection between air and water pollution (both largely chemical processes) on one hand and urban sociology on the other, and radiocarbon dating has been a most important tool to archaeologists for estimating the ages of objects of historical interest.

Chemistry is a largely quantitative science dealing with exact numerical measurements as in chemical formulas and equations. It is impossible to eliminate this quantification and still have a study that is recognizable as chemistry.

WHAT IS THE SCIENTIFIC METHOD?

A scientist may admire a sunset for its aesthetic values, marveling at its beauty, and not for a moment attempt to analyze its causes. At other times, the sunset might be regarded as the scattering of the longer wavelengths of visible light by colloidally dispersed water droplets and solid particulate matter in the atmosphere. In light of this, one might ask whether the acquisition of scientific knowledge about sunsets diminishes the mystery associated with them. Does it not involve some sort of loss of innocence? The answer, fortunately, is probably "no" in most instances. The reason is that we tend to compartmentalize our thinking; we are scientists for part of the time and ordinary people for the rest. And after all, aren't the aesthetic and scientific ways of looking at a sunset merely different aspects of the same truth? Each way portrays a sunset correctly, but does so from entirely different frames of reference. Neither by itself can provide a complete description; they both complement each other to make up the whole.

The term "scientific method" is frequently applied to the working behavior of scientists, designating their approach to solving scientific mysteries. The use of this term is unfortunate, because it conveys the impression that there is a tried and true "method" that will provide the answers to all scientific questions. Since each research problem is different, it stands to reason that the "method" used for each solution will vary accordingly.

Perhaps a better term is "scientific attitude." This refers to the state of mind which all scientists should have before setting foot inside the laboratory. First of all, the scientist should possess a high level of motivation. One important motivational influence is simply curiosity. To want to know what lies on the other side of the hill or around the bend in the road or on the far side of the moon or within the nucleus of an atom is fundamental to human nature. It is part of our quest for ultimate values and answers. But rarely is a scientist motivated by curiosity alone. There may be humanitarian reasons, such as seeking a cure for cancer or developing a new strain of corn, rich in proteins, for use in countries where the diet is normally deficient in protein. The desire for personal fame and recognition is often a factor. Or a college professor may simply want a promotion or academic tenure and feels that one way to do it is to become the architect of a scientific breakthrough.

A healthy skepticism about experimental findings is a third important attribute. It is a measure of humility and self-control and puts a brake on any tendencies to come to hasty conclusions at the first sign of encouraging results.

Finally, a successful scientist has an unwavering desire for truth, regardless of the consequences. We are all aware that the truth sometimes hurts, but we still have a greater or lesser tendency to evade reality when the truth might deflate one's ego or in other ways prove painful. We all wish at times for two plus two to equal five instead of four. But one must never allow this foible to interfere with the interpretation of results obtained in the laboratory or with how these results are presented to the world.

CHEMICAL TERMINOLOGY

We stated earlier that there are 105 different known kinds of atoms. Each of the 105 represents a distinct chemical *element* with its own unique physical and chemical properties. Elemental sulfur, for example, is a yellow solid that your great grandmother may have taken as a tonic, mixed with molasses, many years ago. You are all familiar with elemental iron and its many uses. The oxygen that constitutes about 20% of the air we breathe is a colorless gas essential to all higher forms of life. Elemental chlorine, on the other hand, is a deadly poisonous yellow-green gas that was responsible for hundreds of thousands of casualties on the battlefields of World War I. Each element is denoted by an atomic symbol. The symbol for sulfur is S and the symbol for iron is Fe (after the Latin word for iron, *ferrum*) while those of oxygen and chlorine are O and Cl, respectively.

Substances composed of two or more different elements chemically combined are called *compounds*. Compounds should be distinguished from mere physical mixtures of elements. For example, it is possible to grind finely divided iron together with sulfur to make a physical mixture that to the eye will have a completely uniform appearance. Examination of the mixture under a microscope will show, however, that it still consists of separate particles of iron and sulfur. Furthermore, the two substances can be readily separated with a magnet.

Now take the mixture and strongly heat it. This causes the iron and sulfur to chemically react and form the compound ferrous sulfide (FeS), a black powder. A sample of FeS has a completely homogeneous appearance even under the most powerful microscope and has entirely different physical and chemical properties than a mixture, regardless of how finely divided, of iron and sulfur. This is because in FeS the atoms of iron and sulfur have formed chemical bonds with one another whereas in the physical mixture they have not.

Compounds may be grouped into two broad categories, **molecular compounds** and **ionic compounds**. The former exist as distinct particles, called *molecules*, which contain two or more different chemically combined atoms. An example is water, represented by the formula H_2O. This signifies that every molecule of water contains two atoms of hydrogen and one of oxygen (Fig. 1-4). Another well known molecular compound is sulfuric acid (H_2SO_4), which is found mixed with water in automobile batteries. Each molecule of pure H_2SO_4 contains two atoms of hydrogen, one of sulfur and four of oxygen. Elements also exist as molecules. Some, such as helium (He), possess only a single atom per molecule. Others have two or more atoms, like oxygen (O_2), with two atoms in each molecule, and sulfur (S_8), with eight.

Ionic compounds are made up of *ionic lattices*, which are three dimensional networks of electrically charged atoms called ions. Some ions carry positive charges, and some are negatively charged. The sodium ions (Na^+) bear a single positive charge, the chloride ions (Cl^-) a single negative charge. They are found in equal numbers in the lattice (in order to maintain overall electrical neutrality), as is signified by the formula "NaCl".

Not all ionic lattices have ions present in a 1:1 ratio. Magnesium chloride ($MgCl_2$) has two chloride ions in the lattice for every magnesium ion. This is because it takes two singly charged chloride ions Cl^{-1} to electrically neutralize each doubly charged magnesium ion (Mg^{+2}). More will be said later about ionic compounds.

CHEMICAL FORMULAS

Just about any chemical reaction can be put down in words. For example:

$$\text{hydrogen gas + oxygen gas} \xrightarrow[\text{spark}]{\text{electric}} \text{water}$$

The "+" sign means plus and "→" means yields. The words "electric spark" indicate the necessary condition for the reaction to occur. This word equation tells us that water can be produced from its elements, hydrogen and oxygen. In this equation, the hydrogen and oxygen gas are called the reactants. This means that they are the starting materials. The water is called the product, meaning that it is the substance produced by the reaction.

However, word equations tell us nothing about the chemistry of the reaction. We can remedy this situation by writing the formula of each substance in place of the words.

electric

hydrogen gas + oxygen gas → water

spark

electric

$$H_2 + O_2 \rightarrow H_2O$$

spark

We have written the symbols in place of the words. However, we must perform one additional task and that is to balance the equation. We must be sure that the same number of atoms of each element appear on both sides of the equation, the reactant side and the product side. Chemists have found that atoms are not lost or gained in a chemical reaction, this is what has come to be known as the Law of Conservation of Mass. (We should point out that atoms are not lost or gained in common chemical reactions. In nuclear fission and fusion, atoms can be lost or gained by being partially turned into energy.)

In the equation for the formation of water that we've just written, you will see that there are two hydrogen atoms and two oxygen atoms on the left side of the equation, and two hydrogen atoms and one oxygen atom on the right side of the equation. This equation is unbalanced; there is one less oxygen atom on the right side of the equation as opposed to the left. The way we balance a chemical equation is to use coefficients, or numbers that we place in front of the formulas of elements and compounds. The equation cited below is balanced.

spark

$$2H_2 + O_2 \rightarrow 2H_2O$$

AIR POLLUTION

It is a common misconception that air pollution has become a problem only within the last few decades or so. Actually, it has probably been with us in some form or another since the very beginning of our existence on earth. In fact, not all air pollution is caused by human activities. The haze from which the Great Smoky Mountains derive their name is thought to be caused by evaporation of terpenes and other chemicals from the trees covering the region. Ultraviolet radiation from the sun causes these substances to react with atmospheric oxygen in such a way as to give rise to the finely divided particles responsible for the poor visibility that frequently disappoints tourists driving through that area.

Even man-made pollution is nothing new. Neanderthals must have suffered from its effects when they had to spend long ice age winters in caves filled with smoke from fires built to keep warm and cook food. Other primitive peoples whose dwellings do not provide for efficient disposal of smoke have been afflicted with eye and respiratory ailments. But it is especially since the Industrial Revolution, coincidental with the appearance of William Blake's "dark Satanic mills" in the English Midlands in the early nineteenth century, that industry has begun to affect entire cities and regions. Today, industrial activity is so great and fuel-powered transportation devices so numerous that this industrially-driven air pollution is starting to produce noticeable effects on a global scale.

Which of society's pursuits seem to be the worst offenders with regard to air pollution? Almost half of the pollutants in the United States come from transportation devices. The principal pollutants from motor vehicles are carbon monoxide and the smog producers.

Carbon monoxide (CO) emitted by automobile engines arises from incomplete combustion of the hydrocarbons found in gasolines, in which the fuel reacts with less oxygen than it is theoretically possible. Automobile exhaust gases contain an average of 4–5% carbon monoxide, and during one year's operation of an "average" automobile some 3,200 lbs. are emitted. Although its toxic properties are well known, its lack of color, taste and odor make it especially insidious and dangerous.

Air Pollutants by Source

Transportation	43.5%
Stationary sources (power plants, etc.)	19.7%
Industrial processes	13.9%
Garbage incineration	5.1%
Miscellaneous	17.8%

A running automobile in a closed garage will rapidly build up a lethal concentration of the gas, but even on city streets crowded with traffic, levels are frequently attained that are far in excess of the federally recommended maximum concentration for an eight-hour time-weighted average exposure of 35 parts per million (ppm). Concentrations of 100 ppm and more are not unusual. Individuals such as traffic policemen, taxi drivers and garage mechanics who are continuously exposed to this would appear to have some cause for worry. In fact, tanks of pure oxygen have been placed at busy intersections in some cities for the benefit of traffic policemen. Every half hour or so, they are required to take an "oxygen break" to counter the effect of exposure to CO and other traffic fumes.

Air Pollutants by Type

Carbon monoxide	43.7%
Hydrocarbons	13.5%
Nitrogen oxides	7.2%
Sulfur oxides	13.4%
Particulates (solid particles)	12.3%
Other	9.9%

Carbon monoxide is poisonous because it interferes with transport of oxygen by the blood to the tissues. In the normal course of respiration, oxygen in the lungs is picked up by the hemoglobin of red blood cells to form a loosely bonded product called oxyhemoglobin, which is then carried to cells all over the body. There, the oxygen is released and is used by the cells to metabolize nutrients. When carbon monoxide is inhaled, however, it forms a complex with hemoglobin that is 300 times more stable than oxyhemoglobin. Hemoglobin in this form cannot pick up oxygen, and if enough has been deactivated by the CO, the victim in effect dies from suffocation.

It is estimated that on a worldwide basis, human activities together with various natural phenomena release some 260 million tons of carbon monoxide into the Earth's atmosphere every year. Yet its overall concentration does not seem to be increasing. There is rather convincing evidence that soil bacteria are the agents responsible for its disappearance. Experiments utilized a sealed chamber in which was maintained an atmosphere containing varying concentrations of CO together with some normal soil from the outdoors. In another chamber, the researchers established the same conditions except that they first sterilized the soil with heat, thereby killing all the bacteria in it before placing it in the chamber. The CO concentration in the first chamber dropped rapidly, due to its consumption by the bacteria, while no change was observed in its concentration in the second chamber.

Smog of the type usually associated with Los Angeles was first noticed in that city around 1942, but it took years to discover its true cause. At first, it was thought to arise from smoke and dust emitted from incinerators and factories. Accordingly, the Los Angeles County Pollution Control Board issued a ban on all outdoor burning of trash and initiated steps toward control of industrial smoke emission. But the smog continued unabated, so an accusing finger was next pointed at sulfur dioxide given off by oil refineries in the Long Beach area and by the combustion elsewhere of sulfur-bearing coal. Controls were placed on SO_2 emissions, but still to no avail.

The cause of the problem was finally deduced as the result of a chance discovery by a biochemist, Dr. Arie Haagen-Smit, whose research was aimed at finding the compounds responsible for the pleasant tastes and odors of fruit. One day, he was attempting to isolate the principle of pineapple odor by drawing air through a preparation of the crushed fruit and passing it through a trap, cooled by liquid nitrogen, in which the essence would condense. But when he removed the trap after the run was over, he detected an unmistakable odor of ozone, a highly toxic and reactive material that one would hardly expect to find in pineapple. Noticing an unusually high level of smog that day, he acted on a hunch and pulled a quantity of ordinary air through the trap. He obtained a murky liquid with a very unpleasant odor. Separation of the mixture into its components showed it to consist of a mixture of hydrocarbons, aldehydes, organic acids, organic nitrates and ozone. Where could these substances have been coming from? Later investigations by him and by others showed beyond the shadow of a doubt that the automobile was the principal source.

Just how do motor vehicles produce smog? Some of the details are not yet clear, but the following, in somewhat simplified outline, appears to be what happens. First, the high temperatures in the engine cause atmospheric oxygen and nitrogen to react, producing nitric oxide:

$$N_2 + O_2 \rightarrow 2NO$$

At the same time, varying quantities of the fuel in the piston chamber fail to burn completely with the result that a mixture of olefins, aldehydes, ketones and aromatic hydrocarbons are expelled from the exhaust and crankcase. There is also a certain amount of simple evaporation of fuel from the gas tank and carburetor. Ultraviolet radiation from the sun now enters the picture, causing a complex series of reactions involving the nitric oxide, atmospheric oxygen and the aforementioned organic compounds. Nitrogen dioxide (NO_2) and ozone are formed, both highly irritating and toxic substances. Fortunately, their concentrations in outdoor air never reach high enough levels to give rise to unmistakable symptoms of poisoning in human beings. The same, unfortunately, cannot be said about another constituent of smog, peroxyacetyl nitrate (PAN):

$$CH_3COONO_2$$

PAN is a powerful lacrimator (tear producer) and respiratory tract irritant. High school sports events in the Los Angeles area have been postponed on numerous occasions during periods of high smog concentration.

Smog is by no means limited to Los Angeles. Tokyo had an especially severe episode in June, 1970. It got so bad that oxygen tanks were set up on street corners to assist those with breathing difficulties. Even mile-high Denver, in a region supposedly noted for the purity of its air, has had its share of days on which people would rub their smarting eyes.

Certain weather conditions seem to favor smog formation. One prerequisite is plenty of sunshine so that the necessary photochemical reactions can take place. There is no lack of that in Los Angeles during its smog season from April to November. Furthermore, there is little precipitation at that time. Most of Los Angeles' yearly total of 15 inches falls in the winter months. Nothing cleans the air as well as a good rainstorm.

Another requirement is *temperature inversion*. This happens when relatively cool air becomes trapped beneath a layer of warm air. Pollutants tend to accumulate in the lower inversion layer and are prevented from dispersing by the warm air, which acts as a "lid." In this manner, the pollution builds up to uncomfortably high levels.

WATER POLLUTION

Life as we know it is inconceivable in the absence of water. It is essential that all living organisms, including humans, have access to usable sources of this substance. The human body is 70% water by weight, and one of its principal functions there is to act as a neutral, non-toxic solvent within which biochemical reactions may take place. Water is also instrumental in the elimination of waste products such as those found in urine; these must be transported via the bloodstream to the kidneys where they are separated through the dialysis process and excreted. Water is an actual participant in many biochemical transformations. Digestion of food is primarily a series of hydrolyses (reactions with water) of such substances as proteins, fats and carbohydrates so as to break them down into simpler materials that can then be absorbed through the walls of the small intestine and into the blood-stream. Finally, water plays a vital role in the regulation of body temperature of birds and mammals. Because of its high specific heat (ability to absorb heat with a minimum of temperature change), it is eminently suited for absorbing the heat evolved by the numerous biochemical reactions of the body, thereby facilitating a constant body temperature.

Unfortunately, and owing in large measure to pollution, it has become impossible to drink water in many areas of the world without expensive water treatment facilities. The water pollution problem in the United States is to some degree due to indifference and the desire to minimize costs on the part of some industries. It is also partly the result of inadequate sewage disposal procedures in many cities and towns. In a number of areas, water pollution laws are insufficiently broad in their scope, are poorly enforced or both.

That form of water pollution posing the most direct menace to human health is *bacteriological contamination*. Historically four principal water-borne diseases, all transmitted by microorganisms from human wastes are: cholera, dysentery, typhoid and hepatitis.

Water pollution by household and industrial detergents has also been a major worry in recent years. One aspect of it that, until a few years ago, posed a serious problem was that certain commercial detergents possessed chemical structures that were not easily degraded by the bacteria found in rivers, streams and lakes. The detergent content of the water would continue to build up until it reached undesirable high levels. "Foaming" became a nuisance in sewage treatment and water purification plants. Sometimes sudsy water even issued from household faucets. Instances of drowning of whole flocks of wild ducks that happened to land on detergent-saturated ponds were reported. The detergent, acting as both a wetting and emulsifying agent, washed the oils from the ducks' feathers which allow the ducks to float.

The inability of these early detergents to be degraded was caused by branched hydrocarbon side chains that were not readily attacked by microorganisms. In 1965, major detergent manufacturers switched to biodegradable detergents that possessed straight-chain hydrocarbon groups that bacteria could metabolize. Since then, that aspect of the detergent pollution problem has largely disappeared.

Another troublesome factor, however, remained unresolved. It arose because in order for most detergents to display maximum cleaning effectiveness, it was necessary to incorporate inorganic salts, called *builders*, into the detergent formulations. Perhaps the most widely used builder was sodium tripolyphosphate ($Na_5P_3O_{10}$). Phosphates are important nutrients for aquatic plant life (as well as for all other life), so when large quantities of phosphates from detergents or other sources are discharged into bodies of water, an explosive growth of algae, called an *algae bloom*, may take place. The algae, on dying, load the water with organic matter that, through its oxidation by bacteria, uses up dissolved oxygen fish need to survive. The term for this is *eutropification*. Perhaps the most noteworthy example of eutropification is Lake Erie. Even though it is a massive body of water, it has the misfortune of having a number of major metropolitan areas either right on its shores (Cleveland, Buffalo and Erie) or situated on streams that flow directly into it (Detroit and Toledo). Phosphates from the sewers of these cities and nitrogen compounds from fertilizers used on surrounding farmlands have been accumulating in the lake at rates faster than the lake can cleanse itself through its outlet over Niagara Falls. The excessive growth of algae, especially during the summer months, causes such a drop in oxygen level that many useful types of fish (bass, whitefish, lake trout) have disappeared from much of Lake Erie. Left behind are such relatively undesirable species as carp, crappies and suckers that can get by on less oxygen. A once thriving fishing industry in the lake has almost entirely disappeared. Only in recent years, and to the surprise of scientists, has fishing resumed to the lake. The better quality of water can be attributed to the efforts of many to meet environmental regulations and discharge limits on contaminants.

DISPOSAL OF SOLID WASTES

We live in an economy that ceaselessly shouts "Consume!" at us, with the result that we are producing staggering quantities of refuse, and we are running out of close and convenient places to put it. The growth of metropolitan areas not only means a greater trash and garbage-generating population, it also cuts down on the number of nearby available sites where its disposal is possible. Who wants a dump, or an incinerator next to their back yard?

The average person in the U.S. produces 5.32 lbs. per day of trash that is actually collected (no one knows how much more simply ends up as litter). And this figure does not include food wastes (garbage).

The cheapest, but by no means the most satisfactory, way to dispose of trash and garbage is an **open dump**. Dump fires and decomposition of refuse create obnoxious odors and give rise to air pollution. Open dumps maintain large rat populations that also frequently infest neighboring areas. Dumps produce swarms of flies that invade nearby homes, carrying filth directly from the dump to the dinner table. Yet a sizeable percentage of the towns and cities in the U.S. still get rid of their refuse in this manner.

Incineration, if carried out properly with anti-pollution devices, works very well for combustible materials. An obvious disadvantage of even the best incinerators is that they do not dispose of non-combustibles such as cans and bottles. Every effort is made, therefore, to separate as much of this material as possible; otherwise, glass and many metal alloys melt and form a slag at the bottom of the incinerator that is difficult to remove and takes up increasing amounts of space as more and more of it accumulates. Every incineration process also yields a certain amount of ash (non-combustibles) which must be disposed as hazardous waste typically in a *secure landfill*. The secure landfill is designed to contain hazardous wastes, although in recent years many commercial chemicals have been banned by the U.S. EPA from disposal in secure facilities. This was based on a product's toxicity and its mobility in soil and groundwater. Under the Land Ban, as this regulation is called, a wide variety of commercial solvents, cyanides, sulfides and other substances are regulated relative to land disposal. The federal Resource Conservation and Recovery Act (RCRA) provides specifications for secure landfills, including the compaction of the soils utilized; the use and nature of synthetic liners; the collection of leachate generated as rainfall percolates through the soil absorbing to some degree soluble materials buried in the landfill, as well as carrying along particles of soil on which other chemicals are absorbed. These must be detailed in a RCRA Part B permit application which is submitted to and must be approved by the state environmental agency as well as U.S. EPA.

The *sanitary landfill* technique is often utilized for disposal of refuse. In one such procedure, a trench approximately eight feet deep is first dug in the ground. The waste is dumped in and compacted. Intermittent layers are usually covered with dirt. When the trench has been filled to within about two to three feet of the top, dirt from another trench is piled in to cover the refuse. The landfill method is less expensive than incineration and permits disposal of all kinds of waste. The rats, flies and obnoxious odors associated with open dumping are eliminated. It does, however, have certain disadvantages. In the first place, land for this purpose may not be available. Furthermore, landfill sites, while they may be made into parks, golf courses and other recreational areas, should not be used for construction sites. Explosive methane gas is produced by bacterial action on the buried refuse, and buildings constructed on the fill may accumulate pockets of it. A number of explosions in buildings situated on former landfill sites have been recorded.

Some ingenious processes have been developed to provide ways of obtaining useful products from refuse. The U.S. Bureau of Mines Coal Research Center in Pittsburgh has worked out a procedure that converts garbage into a petroleum-like material with good fuel properties. The garbage, which contains about 40% by weight of organic material (largely plant fibers or cellulose) and 60% water, is placed inside a thick-walled steel pressure reactor. The vessel is heated to 380°C, and carbon monoxide is introduced at a pressure of 1,000 pounds per square inch. The following reaction takes place between the carbon monoxide and the steam generated inside the reactor:

$$CO + H_2O \xrightarrow{\text{steam}} CO_2 + 2H \text{ (nascent hydrogen)}$$

The *nascent*, or *atomic* hydrogen, being a very reactive substance, combines with the organic material in the garbage and converts it into a dark liquid that, like the petroleum it resembles in appearance, consists largely of combustible hydrocarbons. Experimental results show that the process can yield about 500 lb. of "petroleum" from one ton of garbage. At first blush, it seems difficult to believe that the "petroleum" produced in this manner could compete in terms of cost with its counterpart that is obtained from the ground, but that is beside the point. Our first priority here is not to manufacture petroleum; we should look upon the oil merely as an extra dividend. The overriding concern is to get rid of the garbage in an ecologically safe manner; and regardless of how we do it, disposal is going to cost money.

THE CONCEPT OF RECYCLING

Even though the word "recycle" has become a generally familiar term to us, the idea it represents has been in operation on this planet ever since life here made its first appearance. In the absence of pathways for recycling of matter, any organisms that might have initially come into being would have died out as soon as existing supplies of nutrients had become exhausted. The material forming the nutrients would, due to the action of living organisms, have ultimately resulted in the formation of simple low energy substances such as CO_2, H_2O and N_2. In order for life to continue, it was necessary for nature to provide some means for changing these simple substances back into high energy nutrients. A key step in such natural recycling is *photosynthesis* in which green plants transform CO_2 and H_2O into carbohydrates with the aid of radiant energy from the sun. In fact, the sun, through photosynthesis, is the ultimate source of all the energy required by life. Another indispensable part of the cycle of life is *nitrogen fixation* by bacteria; through it, atmospheric nitrogen is changed into forms required for making proteins, nucleic acids and other biologically important molecules. Eventually all biological matter returns, through metabolism of nutrients, excretion of wastes or the death of the organism, to the simple low energy molecules from which they arose, and the cycle begins again. The purpose of such natural recycling is to provide for the reuse of matter over and over again in the biosphere while at the same time making energy available to animals and plants for their vital functions.

Unlike natural phenomena, mankind's methods of utilizing matter and energy are frequently open-ended and make no provisions for the reutilization of material. It is becoming increasingly evident that such a course of action can lead nowhere but to disaster, for the following two reasons that should have been evident long ago: (1) **Natural resources are being consumed at a continuously accelerating rate**, with no provisions being made for their replacement or reuse, and (2) **Enormous quantities of man-made wastes are entering the environment**, forming cycles of their own that frequently disturb natural cycles. The assumption has traditionally been that our air, hydrosphere and available land are limitless and that somehow, wastes dumped into them would magically disappear.

Our economic system has long been geared to the concept of maximum profit. Unless recycling can be shown to yield a satisfactory financial return, it has been virtually a foregone conclusion that it will never be voluntarily tried. However, aroused public opinion and environmental legislation has changed this picture. Necessary changes in attitude and practice are being applied with success. Voluntary pollution reduction programs proposed by EPA Administrator William F. Reilly have been very successful. Mr. Reilly's 33/50 Reduction Program has amazed critics and supporters. The recent Environmental Conference in Brazil also points out a global concern with environmental and recycling issues.

Fortunately, there are some areas where, according to traditional profit-oriented economics, recycling yields a return. "Nonreturnable" bottles are now frequently being returned to the manufacturer and remelted; in numerous communities there are collection sites where families can leave their week's supply of empty glass containers. Scrap iron now furnishes almost as much iron to the steel industry, in any given year, as ore mined from the ground. An estimated 79% of all junked automobiles are recycled as scrap.

The chemical industry has many processes in which "waste" materials are recycled or converted into useful by-products. This recycling is not just a recent phenomenon. The *Solvay process*, the manufacture of two important industrial heavy chemicals, sodium carbonate and sodium bicarbonate, was developed in Belgium around 1865. Ammonia and carbon dioxide are first added under pressure to a solution of sodium chloride, and the following reaction takes place:

$$NaCl + NH_3 + CO_2 + H_2O \rightarrow NaHCO_3 \downarrow + NH_4Cl.$$

The CO_2 used in the process is made by heating limestone at 900 °C (Δ means "heat"):

$$CaCO_3 \xrightarrow{\Delta} CaO + CO_2 \uparrow$$

while the ammonia is made by the *Haber process*.

The sodium bicarbonate which precipitates from the reaction mixture is converted into sodium carbonate by heating, which drives off CO_2 and water:

$$2NaHCO_3 \xrightarrow{\Delta} Na_2CO_3 + H_2O + CO_2 \uparrow$$

The CO_2 produced in this manner ($\uparrow$ means "in the vapor or gaseous state") is recycled and used again in the first step, along with CO_2 obtained from limestone. Meanwhile, the NH_3 (ammonia) is regenerated by treating the NH_4Cl (ammonium chloride) produced in the first step with the CaO that came from the heating of the limestone:

$$CaO + 2NH_4Cl \rightarrow 2NH_3 \uparrow + CaCl_2 + H_2O$$

The only troublesome product from an environmental viewpoint in the whole process is the calcium chloride, which is often simply dumped into the nearest stream. Because it is *hygroscopic* (water absorbing), some $CaCl_2$ is sprinkled on dirt roads to keep the dust down. It is also used for melting ice and snow on highways during the winter. But even these uses are less than good environmental practices, since the Ca^{++} and the Cl^- ions, after being washed off the roads, frequently come to contaminate ground water.

The recycling of commercial solvents such as acetone, paint thinner (petroleum naphthas, toluene, xylene), trichloroethylene (a suspect carcinogen) and perchlorothylene (dry-cleaning solvent) is occurring more frequently among chemical remanufacturing and commercial solvent-using industries. This process entails controlled distillation, by which the waste solvent liquids are placed into an enclosure (or pot), heated to boiling and the resultant purified vapors condensed (reliquified) and collected in a separate container.

The condensate (condensed vapors of solvent) which has been purified in the still is then cooled to room temperature and can be used inside the plant or shipped as recycled solvent (not waste) to another user or recycling company. The recycler may then blend various solvents to generate a solvent or solvents which meets his buyers' specifications. This process significantly reduces waste solvent volume, thereby minimizing handling and disposal costs. In addition, the recycling of waste solvent generates a usable product, potentially reducing the plant's raw materials (solvents) budget. This process is not totally devoid of wastes; high-boiling impurities (perhaps 15 to 20% of the original waste solvent volume) do hang back in the distillation pot and must be disposed of as hazardous waste, usually by incineration due to the frequently toxic and/or flammable nature of the solvent.

ADDITIONAL REFERENCES:

Hall, Ridgway M., Jr. et al., *RCRA Hazardous Wastes Handbook, 9th Edition*, Government Institutes, Rockville, MD, 1991.

Bauer, Mary and Elizabeth Kellar, *Managing Your Hazardous Waste: A Step-by-Step Guide*, Government Institutes, Rockville, MD, 1992.

Chapter 3

• • •

BASIC AIR QUALITY

James W. Little
Dames & Moore

OVERVIEW

The air we breathe has been a factor in the location of villages and towns not long after the first settlements. With the Industrial Revolution, poor air quality became first recognized as a health issue and for those areas with naturally stagnated air, the issue became serious in the 19th Century. Major European cities such as London and Paris had air quality problems from factories and fireplaces. Interestingly, during that same time, the Los Angeles Basin was known as the "Valley of the Smokes" from the campfires and Indian settlements. This early warning about adverse climactic conditions did not stop settlements but has today resulted in Los Angeles and California legislating some of the most restrictive air pollution requirements.

To have good air quality is a plus for cities attracting industry, as well as a factor for people choosing a place to live and raise a family. Advertisements push "fresh air" and being pollution free.

Air quality is impacted by those things we can see by eye or under the microscope (dust, pollen, etc.), and those substances we can't see (sulfur dioxide, ozone, etc.). Standards have been developed for many of these compounds, with new compounds regulated each year. Today, we daily learn of some type of "air quality index" on the TV news, on the radio, or in the newspaper.

Air pollutants in the atmosphere cause great concern because of potential adverse effects on human health. Adverse human health effects attributable to air pollution include acute conditions such as respiratory difficulties and long-term effects such as cancer formation. Other adverse impacts of air pollution can also occur, including vegetation damage, damage to materials, and visibility degradation.

TYPES OF AIR POLLUTANTS

Air pollutants can be grouped in different categories. A basic classification approach is to consider air pollutants as either in gaseous or particulate form. Common gaseous pollutants are carbon monoxide, sulfur dioxide, nitrogen dioxide, and ozone. The particulate matter found in the atmosphere can be made up of many different compounds including mineral, metallic, and organic compounds.

Another important distinction is the difference between primary and secondary air pollutants. Primary pollutants are those that are directly emitted to the atmosphere. A common example is the carbon monoxide emitted from automobile exhausts. Secondary pollutants, on the other hand, are formed in the atmosphere as the result of various transformation mechanisms involving primary pollutants or other secondary pollutants. For example, one of the pollutants of most concern in both urban and rural environments is ozone. Ozone is a secondary pollutant formed from the photochemical reaction of volatile organic compounds and oxides of nitrogen. The substances that react to form ozone and other secondary pollutants are referred to as precursors.

Another term used in the categorization of pollutants is the term *regulated pollutants*. Many hundreds of pollutants can occur in the earth's atmosphere, but not all potential pollutants are specifically addressed in air pollution laws and regulations. These pollutants that have been singled out for regulatory control are sometimes referred to as "regulated pollutants". In the United States, this term primarily applies to pollutants identified in the federal Clean Air Act and in the regulations adopted to carry out this act.

Although we often think of air pollution as arising from human activities, many natural sources of air pollution also exist. Natural forms of pollution include wind-blown dust from barren areas, organic emissions from living vegetation, emissions resulting from biological decay, hydrocarbon emissions from petroleum seeps, and emissions from volcanic eruptions.

TYPES OF AIR POLLUTION SOURCES

Mobile and stationary sources are the two basic types of air pollution emissions. Mobile source emissions refer to the emissions generated by the combustion of fuels in various types of motor vehicles (automobiles, trucks, buses, motorcycles, etc.). The principal pollutants emitted from the operation of motor vehicles are carbon monoxide, oxides of nitrogen (primarily nitric oxide and nitrogen dioxide), and hydrocarbons. Vehicle emissions are of greatest concern in urban areas where the volume of traffic is highest and where vehicles often operate under conditions (such as stop-and-start driving) that result in higher emission rates.

Stationary sources are often classified as either point sources or fugitive sources. Point sources are those that emit air pollutants through a confined vent or stack. The stack of a fuel-fired boiler is a common example. Fugitive emissions, on the other hand, are those emissions that enter the atmosphere from an unconfined area. Fugitive emissions are often thought of as dust or particle emissions, but the term fugitive can equally well apply to many types of gaseous emissions. Examples of fugitive emissions are the dust particles stirred up by vehicles moving on unpaved roads and the vapors released from the unconfined application of paints containing organic solvents.

Stationary source emissions at industrial facilities can result from fuel combustion and from various process operations. Major industrial activities and categories resulting in air pollution emissions include the following (with examples in parentheses):

- external combustion sources (combustion of fossil fuels)
- solid waste disposal (waste and sludge incineration)
- stationary internal combustion sources (gas turbines)
- evaporation loss sources (surface coating, organic liquid storage)
- chemical processes (synthetic organic chemical manufacturing)
- food and agricultural industry (chemical fertilizer production)
- metallurgical processing (smelting operations, iron and steel production)

- mineral products industry (cement manufacturing, aggregate quarries)

- petroleum industry (refineries)

- wood products manufacturing (pulp and paper mills)

ATMOSPHERIC DISPERSION, TRANSFORMATION, AND DEPLETION MECHANISMS

Pollutants released from both stationary and mobile sources are subject to atmospheric dispersion, transformation, and depletion mechanisms. These concepts are important in understanding basic air quality issues.

Dispersion—The dispersion of pollutants in the atmosphere is determined by mean wind flow conditions and by atmospheric turbulence. Turbulence results from such factors as the friction of the earth's surface; physical obstacles to wind flow including man-made structures and natural terrain features; and the convective conditions of the atmosphere. Some of the most important aspects of air pollutant dispersion are the following: Stability Class; Wind Changes With Height; Inversions and Stagnation Episodes; Emission Source Dispersion Characteristics and Plume Rise.

- *Stability Class*—Stability class refers to the degree of turbulence existing in the atmosphere as the result of wind speed and convective conditions related to the change of temperature with height above the earth's surface. Moreover, the stability of the atmosphere usually refers to the lower boundary of the earth where the pollutants are emitted. The idea of discrete stability classes is, of course, a simplification of the complex nature of the atmosphere, but has proved useful in predictive studies. (1) A *stable* atmosphere is marked by air that is cooler at the ground than aloft, by low wind speeds, and consequently, by a low degree of turbulence. A pollutant plume released into a stable lower layer of the atmosphere can remain relatively intact for long distances. (2) An *unstable* atmosphere, on the other hand, is marked by a high degree of turbulence. A visible plume released into an unstable atmosphere may exhibit a characteristic looping appearance produced by turbulent eddies. (3) An intermediate turbulence class between stable and unstable conditions is the *neutral* stability class. A visible plume released into a neutral stability condition may display a coning appearance as the edges of the plume spread out in a V-shape.

- *Wind Changes with Height*—In most atmospheric conditions, wind speeds tend to increase with height above ground level. In addition, wind direction typically changes with height. Wind shears created by changes in speed and direction affect the transport and diffusion of pollutant plumes from emission sources.

- *Inversions and Stagnation Episodes*—The term *inversion* refers to a layer in the atmosphere where temperature increases with height rather than decreasing as is the usual case. This inversion layer serves as a stable "lid" keeping pollutants emitted below the layer from dispersing further upward. Pollutant levels can build up near the surface as a result of this trapping action. A prolonged period of dispersion is sometimes referred to as a "stagnation" episode. High pollution levels, especially "smog" conditions associated with vehicle emission in urban areas, can occur during stagnation episodes.

- *Emission Source Dispersion Characteristics and Plume Rise*—The same emission rate from two different sources can produce very different ground-level concentrations depending on the dispersion characteristics of the two sources. For example, pollutants emitted from a source with a tall stack tend to produce lower ground-level concentrations than pollutants emitted from a source with a short stack. Sources with the same stack height can produce different impacts depending on the plume rise above the stack. Plume rise is determined primarily by (1) the vertical exit velocity of the exhaust gases and by (2) the temperature of the exhaust gases. The combination of the physical stack height and plume rise above the stack is referred to as the effective stack height.

Transformation—The conversion of precursor substances to form a secondary pollutant such as ozone was previously mentioned. This conversion is an example of a chemical transformation in the atmosphere. Both physical and chemical transformations affect the ultimate impact of originally emitted air pollutants.

Depletion—Pollutants emitted into the atmosphere do not remain there forever. Two common depletion mechanisms are dry deposition and washout. Dry deposition refers to the removal of both particles and gases as they come into contact with the earth's surface. Washout refers to the uptake of particles and gases by water droplets and snow, and their removal from the atmosphere as rain and snow fall to the ground. Acid rain (or, as more generally named, acid deposition) is a form of pollution depletion from the atmosphere.

EMISSION CONTROL METHODS
FOR STATIONARY SOURCES

A variety of methods exist for the control of air pollutant emissions from stationary sources. Among the more important are the following:

Add-On Pollution Control Devices—Add-on control devices remove or destroy pollutants after they are generated but before they are discharged to the atmosphere. Examples of the devices are baghouse filters, electrostatic precipitators, wet scrubbers, carbon adsorption beds, and incinerators. Although often highly effective in removing pollutants, one drawback to this control approach is that frequently a solid or liquid residue is created that must be disposed of in an environmentally acceptable manner.

An important economic consideration in planning an air pollution control program based on add-on pollution control devices is the recognition that control devices typically become progressively more expensive to install and operate as the required control efficiency increases. For example, the added cost of going from 90 percent control efficiency to 99 percent control efficiency could exceed the cost of achieving the initial 90 percent efficiency.

Cleaner Fuels—For stationary combustion emission sources, switching to cleaner fuels can be an effective pollution control method, when technically and economically feasible. Natural gas, for example, is a cleaner burning fuel than most liquid and solid hydrocarbon fuels.

Material Substitution—Some pollution emissions can be reduced by switching to alternative materials. Emissions arising from surface coating operations offer a good example. Considerable progress has been made by the producers of coatings (paints, varnishes, etc.) in reducing the volatile organic solvent content of coatings through use of low-solvent products that still achieve acceptable application, adherence, and longevity properties.

Process Modifications—Like material substitution, process modifications can serve to reduce the original generation of pollution and avoid the need for add-on control devices. In some cases, eliminating a pollution causing process altogether may be more economical than installing control equipment.

REGULATORY APPROACH TO AIR QUALITY

The following discussion summarizes policies in the United States as an example of the regulatory approach to control of air quality. In the United States, the federal Clean Air Act is the basic air pollution law applicable throughout the nation. To implement the Clean Air Act, the U.S. Environmental Protection Agency adopts, revises, and rescinds regulations as needed. In 1990, the Clean Air Act Amendments were passed by Congress adding to the existing Clean Air Act. States and local governments also adopt laws, ordinances, and regulations that, in some cases, may be more stringent than federal requirements.

The key provisions of the federal Clean Air Act (incorporating the Clean Air Act Amendments of 1990) are as follows:

- *National Ambient Air Quality Standards (NAAQS)*—Ambient air quality concentration limits necessary to protect human health and welfare. National ambient standards have been set for sulfur dioxide, nitrogen dioxide, carbon monoxide, ozone, particulate matter, and lead. Nonattainment areas (areas not in attainment with national ambient air quality standards) must take steps to achieve attainment.

- *Prevention of Significant Deterioration (PSD)*—Applicable to attainment areas, a policy for minimizing the incremental increase in pollutant levels above baseline conditions. Proposed "major" new sources and modifications of existing sources in PSD areas must demonstrate that emissions will be controlled with best available control technology.

- *New Source Performance Standards (NSPS)*—Emission limiting standards that must be achieved by designated types of sources.

- *Hazardous Air Pollutants*—A separate list of pollutants, in addition to those regulated by ambient air quality standards. Nearly 200 specific hazardous substances and categories of substances are specified in the Clear Air Act Amendments of 1990. Designated emission sources must apply maximum achievable control technology to comply with national emission standards for hazardous air pollutants.

- *Operating Permit Program*—A program requiring permits for the operation of many emission sources. When it becomes effective, this program will require compliance reports, periodic permit renewal, and annual fees based on the quantity of emissions.

- *Mobile Source Standards*—These standards are requirements for vehicle emission control equipment and for motor fuel characteristics.

- *Acid Deposition Control Requirements*—Procedures for the reduction of sulfur dioxide and nitrogen oxides from power plants. These requirements include the concept of an "allowance" program for sulfur dioxide emissions.

- *Stratospheric Ozone Protection*—A program to implement a phase-out in the manufacture and use of chemicals that cause depletion of ozone in the stratosphere.

AMBIENT AIR QUALITY EVALUATION METHODS

The most widely used ambient air quality evaluation methods are monitoring and modeling. *Monitoring* involves the use of measuring devices to determine the concentration of a specific pollutant, at a specific location, at a specific time. The chief advantage of monitoring is that an exact level of pollution can be determined, subject only to the accuracy and level of detection limitations of the measurement method. The chief disadvantages are: (1) the expense of monitoring prohibits taking measurements at more than a few locations; (2) the contribution of specific emission sources to the total measured concentration can be difficult if not impossible to determine, and; (3) the impact of proposed sources not yet in operation can not be assessed with monitoring methods.

Modeling refers to the use of mathematical representations of pollution dispersion and transformation to estimate ambient pollutant concentrations. The chief advantage of modeling is that modeling can be used to: (1) estimate concentrations at many hundreds of locations at very low overall cost (through use of computer programs); (2) to evaluate the contribution of specific emission sources; and (3) to predict the impact of proposed new sources not yet in operation. The chief disadvantages of modeling lie in the fact that a mathematical model can not exactly replicate the complexities of atmospheric dispersion and transformation. Models, therefore, have inherent accuracy limitations. These limitations tend to be greatest when using models to estimate concentrations at considerable distances from emission points, and when using models to estimate pollutant levels resulting from physical/chemical transformation and depletion mechanisms.

Modeling by sophisticated computer techniques is only as valid as the quality of input data. To be effective as a predictive tool, a model needs to be tested with actual data over an appropriate period of time (model validation).

ADDITIONAL REFERENCES:

Clean Air Act, 42 U.S.C. 7401 et seq. (including the Clean Air Act Amendments of 1990, P.L. 101-549).

Journal of The Air & Waste Management Association (formerly, Journal of the Air Pollution Control Association). Published monthly by the Air & Waste Management Association, Three Gateway Center, Four West, Pittsburgh, PA 15222.

Fundamentals of Air Pollution, Stern, A.C. and Wohlers, H.C., Academic Press, New York, 1984.

Stationary Point and Area Sources, Fourth Edition (through Supplement D, September 1991), U.S. Environmental Protection Agency. 1991.

Compilation of Air Pollutant Emission Factors; Volume I: Office of Air Quality Planning and Standards, Publication No. AP 42. [Available through the Government Printing Office.]

Guidelines on Air Quality Models (Revised), Office of Air Quality Planning and Standards, U.S. Environmental Protection Agency, 1986.

Brownell, F. William and Lee B. Zeugin, Clean *Air Handbook*, Government Institutes, Rockville, MD, 1991.

Building Air Quality, Government Institutes, Rockville, MD, 1992.

Chapter 4

• • •

BASIC TOXICOLOGY

Thomas M. Covilli, C.I.H.
Dames & Moore

OVERVIEW

Toxicology may be defined as the study of the nature and action of poisons—the study of actions of chemicals in the body in order to determine safe levels of exposure and to predict signs and symptoms indicative of excessive exposure. This field incorporates many disciplines such as chemistry, biochemistry, animal physiology and pathology, industrial hygiene, immunology, physics, and statistics.

The field of toxicology can be subdivided into primarily two categories: (1) human, and; (2) environmental. This field is an area of particular focus in recent years by regulatory agencies such as the EPA (Environmental Protection Agency) and OSHA (Occupational Safety and Health Administration). The wide variety of environmental contaminants that are present in the air, soil, and water combined with the range of effects that may result from varying degrees of exposure (via inhalation, skin contact, and ingestion) make this field particularly complex.

For this reason, and due to the fact that it is in its infancy as a field of study, toxicology does not have clear-cut lines of delineation. That is, although there is a substantial amount of objective data regarding animal studies and observed data on human effects from toxic chemicals, it has been difficult for scientists and regulators alike to agree or come to terms with any one set of protocol that clearly and concisely addresses this area.

Consequently, the federal regulators, such as the EPA and OSHA, have, over the past 20 years, developed various standards, regulations, and guidelines which address this area in a significant number of ways. The remainder of this discussion will address the more intrinsic issues of toxicology in an effort to provide a better understanding of this field and how it impacts both industry and the human population in general.

HUMAN TOXICOLOGY

As previously mentioned, toxicology is the science dealing with poisons. Poisons, by definition, are substances that are capable of inducing toxic effects on living organisms. That is, they are capable of causing damage to living tissue, impairment of the central nervous system, severe illness, or, in extreme cases, death when ingested, inhaled, or absorbed by the skin. This should not be confused with the term hazard, which is the *likelihood* that a substance will cause injury under circumstances of normal use. The amounts or concentrations required to produce these results vary widely with the type of substance and the duration of exposure.

Classes Of Toxic Substances

Toxic or harmful substances, in general, can be classified in the following manner according to their physical state:

Dusts. Solid particles of various sizes produced by handling, crushing, grinding, rapid impaction, detonation, of organic or inorganic materials, i.e., rocks, ores, metals, coal, wood, grain, etc. Dusts are not readily diffused in air, and settle under the effect of gravity.

Fumes. Solid particles produced by condensation of volatilized molten metals. Fumes flocculate and may coalesce (i.e., particles come together and cling to one another). One example of a fume is lead fume that is produced during some types of welding.

Mists. Suspended fluid droplets produced by condensation from the gaseous to the liquid phase or by splashing, foaming, or atomizing of a fluid into the gaseous phase. One example of a mist is a typical spray painting operation during which the paint is atomized into a spray.

Vapors. Gaseous form of compounds that are normally either liquids or solids. Vapors can be converted to these states by decreasing the temperature or increasing the pressure. Vapors readily diffuse in air and do not settle readily under the effect of gravity. Two examples of a vapor would be steam or paint solvent vapors.

Gases. Formless fluids occupying space that can be converted to liquid or solids by decreasing the temperature and/or increasing the pressure. They readily diffuse in air. Common examples of gases include nitrogen and oxygen which are present in the air we breathe.

The above classification is restrictive in that it does not cover all categories of substances that may be harmful, and it excludes physical agents such as ionizing radiation (e.g., X-rays and gamma rays) and others which cause body changes through physical means. Also excluded are bacterial molds, parasitic agents, and fungi, which must also be considered as health hazards.

The above categories describe the physical state that a toxic substance may constitute. We will now look at toxic substances according to the type of health risk they pose. Toxic substances can cause a variety of adverse health effects based on their chemical properties and mode of action in the body. The following are generally recognized classifications of toxic substances based on how they effect humans or animals.

Irritants cause inflammation or damage to skin or mucous membranes (e.g., eyes) and may also increase breathing resistance. Irritants can be classified as primary and secondary. Primary irritants form nontoxic end-products and typically do not affect other parts of the body. Examples include many types of acids, such as hydrochloric acid and also substances such as sulfur dioxide which is a component of "acid rain" and finally ammonia gas. Secondary irritants produce toxic effects to other parts of the body, in addition to initial skin or mucous membrane irritation. Examples include nitrogen dioxide, phosgene (nerve gas), hydrogen sulfide and chlorine gas.

The second class of toxic substances based on their health effects is *asphyxiants*. The basis for this classification is that the substance has ability to, either chemically or physically, deprive the body of oxygen.

Simple asphyxiants are physiologically inert gases which act by accumulating in sufficient quantity to exclude an adequate oxygen supply. This type of toxicity normally does not represent a hazard to humans unless it is associated with confined space or vessel entry. Common examples of simple asphyxiants include acetylene, hydrogen, carbon dioxide, nitrogen, methane, and helium. *Chemical asphyxiants* are substances that are able, through their individual or unique toxic action, to render the body chemically incapable of utilizing oxygen. Carbon monoxide, for example, combines with hemoglobin in the blood to form carboxyhemoglobin which then prevents the hemoglobin from transporting oxygen to the parts of the body that need it. Hydrogen cyanide is another type of chemical asphyxiant which prevents certain cell enzymes from using oxygen.

The third classification of toxic substances is *Anesthetics*. These substances are central nervous system depressants and cause narcosis, or sleepiness. Many types of flammable petroleum-derived solvents have this effect, such as toluene and xylene (typical components in paint solvent). Alcoholic beverages are also classified as anesthetics. Most adults can relate with the narcotic effect that excessive alcohol can have on the body. Ethyl ether (ether) is another example of an anesthetic which, at one time, was used to anesthetize patients for surgery.

The fourth classification is *Hepatotoxic* agents which are substances that adversely affect liver function. Common examples of hepatotoxins include a class of compounds known as chlorinated hydrocarbons. Carbon tetrachloride is one type of chlorinated hydrocarbon which was once used as a fire extinguishing media until its effect on the body was known. Perchloroethylene is another type of chlorinated hydrocarbon which is routinely used as a fabric cleaning solvent in dry cleaning facilities. Finally, methylene chloride is yet another example of a chlorinated hydrocarbon which is commonly used in industry as a paint stripper and a degreasing agent.

The fifth class of toxic substances, *Nephrotoxins*, include substances which adversely affect kidney function. The chlorinated hydrocarbons which adversely affect the liver are also often toxic to kidneys. Other examples of kidney toxins include lead compounds, mercury, and turpentine.

Neurotoxins, the sixth classification, include substances that damage the nervous system or interfere with its proper function. Some of the heavy metals, particularly mercury, lead, and manganese, can be particularly damaging to the nervous system. Another example includes a class of substances known as organophosphorous pesticides.

Hemolytic agents comprise the seventh class of toxic substances. These are substances which damage blood or the hematopoietic system (blood-forming organs). A good example of a hemolytic agent is benzene, which was at one time, commonly used as an industrial solvent. Benzene is most often associated with its direct link as a leukemia-causing agent (cancer of the blood-forming system). Other substances which are toxic to the blood include aniline, arsine gas, naphthalene, and nitrobenzene. Nitrobenzene is a common ingredient in shoe polish.

Pulmonary toxins are substances which damage the lung. Substances in this classification cause fibrotic changes (i.e., scarring of lung tissue) over time, and not the rapid irritant action associated with primary irritants discussed earlier. Chronic inhalation of asbestos fibers, for example, has been linked with such disabling lung diseases as asbestosis, mesothelioma (cancer of the lining of the chest cavity), and lung cancer while long term inhalation of silica dust (a component of sand) can cause the disabling lung disease called silicosis. Finally, many coal miners developed a debilitating lung condition known as "black lung" from inhalation of coal dust.

Sensitizers are substances which cause an allergic reaction in sensitized individuals after initial exposure. An individual undergoing a reaction responds to extremely small concentrations after initial exposure. Symptoms typically include constriction of the breathing passages, subsequent breathing difficulty, and skin reactions, such as a rash. A class of compounds known as isocyanates, which are used in urethane paint systems and in the production of foam cushions, is a good example. It was this class of toxic substances that led to many of the deaths in the Bhopal, India disaster. Some sulfur-bearing compounds, such as sulfur dioxide (in acid rain), can also cause an allergic reaction in sensitized individuals.

Carcinogens is a classification of toxic substances which include any substances which stimulate or speed the development of cancer. Cancer is characterized by uncontrolled growth of abnormal cells. Uncontrolled cell division characterizes the disease, frequently a mass of tissue called a tumor. A *direct acting carcinogen* usually causes cancer at the site of exposure. For example, bis(chloromethyl) ether (BCME), when inhaled, causes cancer of the lungs; skin contact with certain coke oven emissions causes skin cancer. *Indirect acting carcinogens* are chemically changed or metabolized by the body into other substances.

Benzidine compounds, for instance, may enter the body through the skin or through the lungs, but they do not cause lung or skin cancer. Instead, metabolites of these compounds are eliminated from the body in urine. These metabolites may cause changes in the lining of the urinary bladder which can result in bladder tumors and bladder cancer. Some examples of other known carcinogens include asbestos (lung cancer and mesothelioma), coal tar pitch volatiles (skin cancer), vinyl chloride (liver cancer), benzene (leukemia), and certain types of chromium compounds (cancer of the nasal septum and lung cancer).

Mutagens are another class of toxic substances which cause deleterious changes in chromosomes within human cells. Chromosomes are the cellular components which carry the genetic code and which make each of us unique. Mutagenic substances have the ability to alter this code, thus affecting human offspring. It has been extremely difficult to prove a direct connection between a chemical exposure and such genetic damage. Ionizing radiation, such as the type released from the A-bombs dropped in Japan in 1945, has been linked to such damage.

Teratogens are toxic substances which interfere with normal embryonic/fetal development. The most vulnerable period is during first 8 to 12 weeks of pregnancy. Malformations, which can result from such chemical exposure, are usually not lethal to mother or fetus. Some typical examples of teratogens include ionizing radiation (e.g., X-rays and gamma rays), thalidomide, some steroids, DDT, parathion, malathion, and glycol ethers. Glycol ethers have been common components of many paint systems.

As the previous paragraphs explained, there is a wide variety of effects that toxic substances can have on the body. It is important to bear in mind, however, that some substances can fall under several of the above categories. For example, benzene, which is a known carcinogen, is also a primary skin irritant and can cause defatting of the skin leading to a form of dermatitis. Thus, certain substances can have multiple health effects.

In discussing human toxicology, the type of health effect is not our only concern. We must also evaluate the intensity or severity of the toxic action. Factors which affect the severity of reaction include rate of entry into the blood stream or body, route of entry or exposure, duration of exposure, individual susceptibility (age, sex, health, previous exposures, allergies) and environmental factors (heat, pressures, stress).

This relationship (i.e., measure of severity of toxic effects on humans) is expressed in the scientific literature as dose vs. response relationship. Dosage is the most important factor in determining whether a given chemical will produce a toxic effect. There is a large variation in the toxicity of chemicals, and even water can cause illness under certain circumstances. Chemicals normally regarded as harmless will evoke a toxic response if added to the body in sufficient amount. For example, salt in the feeding formula of a hospital nursery led to infant mortality before it was discovered by the staff. Only the "dose" determines a poison; no substance is completely nontoxic. In general, it is held that there is "some" measurable dose of any substance which the body will tolerate without overt evidence of tissue reaction.

For comparison of toxicities of different substances, the median lethal dose (LD_{50}) is normally used as the yardstick. The LD_{50} for a particular substance is the weight of poison (usually in milligrams, mg) per unit body weight (in kilograms, kg). The LD_{50} is a statistical estimate (based on animal studies) of the amount of substance required to kill 50% of a given population of test animals. Dose is expressed as quantity per unit body weight (for ingestion), skin surface area (for skin contact), and unit volume of respired air (for inhalation) administered or received per unit time. For example, a substance having an oral LD_{50} of 25 indicates that when 25 mg per kg body weight was administered to the animal test species (via ingestion), half (50%) of the animals died. In general, low LD_{50} values signify highly toxic substances while high LD_{50} values signify lesser toxicity.

When discussing the toxic effects of hazardous substances on the human body, two terms are used to express this relationship as it applies to duration of exposure. "Acute" toxicity refers to exposure of short duration, normally measured in seconds, minutes, or hours (as applied to ingestion, it means a single dose). An example of an acute toxic effect would be inhalation of elevated levels of paint solvent vapors causing a headache while spray painting over a 1 hour period of time. "Chronic" toxicity can be divided into two categories: (1) subchronic exposure and, (2) chronic exposure. Subchronic exposure means intermediate exposures between acute and chronic and usually involves exposure over periods up to 90 days. Chronic exposure means exposure of long duration and, at least with regard to dermal (skin) and inhalation, covers prolonged or repeated exposures with durations of days, months, or years. With regard to ingestion, it means repeated doses of the chemical for months or years. An example of chronic exposure effect would be a person who carelessly removes asbestos-containing materials from buildings and is exposed to the toxic asbestos fibers in the process causing lung cancer 25 to 30 years after such repeated exposures.

Toxic effects may also be local or systemic, depending on the area exposed. *Local* exposure affects the nose, eyes, mouth, throat, skin, and various parts of the respiratory and gastrointestinal (digestive) tracts; absorption into the blood does not have to occur. With *systemic* exposure, however, absorption does occur, and the site of damage may be remote from the site of contact. For example, chronic exposure to perchloroethylene vapors (dry cleaning solvent) can cause damage to the liver and kidney even though it is inhaled into the lungs. In some cases, however, both local and systemic damage occurs.

Routes Of Exposure

The intensity of toxic action is a function of the concentration of the toxic agent which reaches the site of action. Routes of exposure influence this heavily. In general, inhalation is the primary route of exposure in industry, followed by skin or eye contact, ingestion, and injection. *Inhalation* involves breathing gases, vapors, fumes, or mists into the respiratory system which are then absorbed into the blood at small air sac clusters within the lung called alveoli.

With regard to skin contact, the hazardous substance comes into direct contact with the skin and may cause effects either at the direct point of contact or may be absorbed into the skin and subsequently into the blood stream. The main determinants of absorption into the blood include solubility of the substance in water and fat (since the skin is comprised primarily of fat cells) and its molecular size. The skin contact route of exposure is the leading cause of occupational disease, but not in severity.

Eye or ocular exposure involves the hazardous substance coming into direct contact with the eye where it can cause eye damage (due to destruction of cells) or simply be absorbed into the blood with no apparent damage.

With regard to the ingestion route of exposure, the hazardous substance is swallowed or ingested and passes directly into the digestive tract where it is eventually absorbed into the blood. In addition to eating in contaminated areas (the principal method by which toxic substances are inadvertently ingested), the respiratory tract also contributes inhaled particles to the digestive tract. Individual responses vary according to how readily a substance is absorbed into the blood from the digestive tract.

With injection, the liquid or solid phase () of the hazardous substance enters directly into the blood stream via breakage of skin or tissue. Although this method of exposure is not common in industry, it can occur through cuts and abrasions when working with or near toxic substances.

Toxic Substance Exposures— Federal Regulations

In 1970, the Williams-Steiger Occupational Safety and Health Act was passed by Congress. This legislation allowed for the development of standards and regulations by the Occupational Safety and Health Administration (OSHA), a division of the United States Department of Labor, which were to provide a safe and healthful working environment for men and women in general industry.

With regard to chemical toxicology, OSHA soon developed a list of hazardous substances along with specific limits of employee exposure. This list became known as the OSHA "Z" list and the airborne exposure limits which all industrial employers were required to comply with are known as permissible exposure limits (PELs). These PELs are typically expressed as a concentration of contaminant in air (in parts per million or milligrams per cubic meter). These limits are generally based on 8-hour time weighted average (TWA) exposure; however, there are certain highly toxic substances which have concentrations that are never to be exceeded (ceiling values). Employee exposures to these substances must be kept below the respective limits either by means of engineering controls (e.g., ventilation systems), work practice controls, or by respiratory protective equipment.

OSHA has also enacted specific standards for certain extremely toxic substances, such as carcinogens or substances which have very high acute or chronic toxicity. Examples of such substances include lead, benzene, formaldehyde, and asbestos. For these substances, OSHA requires much more stringent controls and medical surveillance requirements to exposed employees.

Finally, OSHA has, over the past 10 years, enacted two specific toxic substance standards which had a significant impact on the industrial community. In 1986, the Hazard Communication Standard (29 CFR 1910.1200) went into effect, which is more commonly referred to as the employee "Right-to-Know" standard. This standard requires importers, manufacturers, and employers who use or handle toxic and hazardous substances to evaluate exposure hazards, provide medical monitoring for exposed employees, train employees, and maintain material safety data sheets (MSDSs) for all such substances. Such companies are also required to maintain a formal written program describing how these measures would be implemented. Components of this HAZCOM standard were expanded to include laboratories who handle hazardous chemicals, via the Laboratory Hygiene Standard which went into effect in January, 1991.

In 1989, OSHA enacted the Hazardous Waste Operations and Emergency Response (HAZWOPER) standard (29 CFR 1910.120) which required employers to provide and implement a "health & safety" program which specifically addresses how it will protect employees against toxic substance exposures, provide medical surveillance to exposed employees, characterize sites for toxic substances, and provide employee training for those employees who are potentially exposed to these substances.

ENVIRONMENTAL TOXICOLOGY

This branch of toxicology addresses the effect of toxic substances not only on the human population, but also on the environment in general, including air, soil (surface and subsurface), surface water, and ground water. The concentration of toxic contaminants and how they affect all living organisms which inhabit our environment are of concern under this branch of toxicology. The EPA, for example, must characterize and evaluate known hazardous waste sites where toxic releases to the environment have been observed. Under EPA's National Contingency Plan (40 CFR Part 300), the EPA must provide a means of responding to toxic substance releases to the environment. The EPA compiled a list of uncontrolled toxic waste sites which were prioritized for long term remedial evaluation and responses. This list of sites has become known as the National Priority List (NPL).

The method by which the EPA decides if and when a site has had a significant toxic substance release such that it justifies being placed on this list is by means of the Hazard Ranking System (HRS). Using this method, the EPA evaluates toxic substance release information relative to the air, soil, surface water, and ground water and determines the *relative risk* that the site represents to human health and the environment. The regulations also mandate that when a remedial action results in residual contamination at a site, future reviews must be planned and conducted to assure that human health and the environment continue to be protected.

As part of its effort to meet these and other regulatory requirements, EPA has developed a set of manuals, collectively titled *Risk Assessment Guidance for Superfund* (RAGS). The *Human Health Evaluation Manual* (Volume I) provides guidance for developing health risk information at NPL sites, while the *Environmental Evaluation Manual* (Volume II) provides guidance for environmental assessment at these sites. Guidance in both human health evaluation and environmental assessment is needed so that EPA can fulfill the environmental regulatory requirement to protect human health and the environment. The *Risk Assessment Guidance for Superfund* manuals were developed to be used for NPL sites, although the analytical framework and specific methods described in the manuals may also be applicable to other assessments of hazardous wastes and material.

The goal of the health and environmental evaluation process is to provide a framework for developing objective, quantifiable data necessary to evaluate the impact of toxic substance releases to human health and the environment. Specific objectives of the process are to:

- provide an analysis of baseline risks and help determine the need for action at sites;

- provide a basis for determining levels of chemicals that can remain onsite and still be adequately protective of public health;

- provide a basis for comparing potential health impacts of various remedial alternatives;

- provide a consistent process for evaluating and documenting public health threats at sites; and

- provide a process for evaluating living resources at or near the site requiring protection, evaluating the effect of the sites contaminants on those resources, and determining the effects on those resources of remedial actions.

As mentioned previously, when evaluating the impact that toxic substances may have on humans or the environment, an evaluation must be made of the four pathways: air, soil, surface water, and ground water. For each pathway, a separate evaluation must be made with regard to three categories:

1. Likelihood of release
2. Characteristics of the toxic contaminants
3. Human, animal, and plant target receptors

The likelihood of release addresses such issues as whether or not toxic substances have been detected at all and, if so, what the likelihood of their reaching the target receptors is. The characteristics category addresses issues such as the acute/chronic toxicity of the substances of concern (i.e., carcinogenicity, etc.) and the concentrations or amounts detected. The target receptors category involves evaluating proximity of toxic substances to nearest humans, plants, and animals. Each pathway can then be objectively evaluated based on quantifiable formulas available in the scientific literature.

ADDITIONAL REFERENCES:

Lee, C. C., *Medical Waste Incineration Handbook*, Government Institutes, Rockville, MD, 1990.

Toxicology Handbook, Government Institutes, Rockville, MD, 1986.

Conner, John D. Jr., *TSCA Handbook, 2nd Edition*, Government Institutes, Rockville, MD, 1989.

Chapter 5

• • •

OVERVIEW OF
ENVIRONMENTAL PROCESSES

Robert J. Taylor
Dames & Moore

OVERVIEW

With some basic understanding of the sciences of geology, ground water, air, and toxicology, the proper evaluation of contaminants in the environment is critical to the cleanup of those compounds. The laboratory can provide many compounds in parts per billion (ppb) and these levels are, in some cases, cleanup criteria for the regulatory agencies. It is of little surprise that with these minuscule levels, the techniques of sampling soil and water (and decontamination of sampling equipment) are very critical.

Through improper disposal or discharge, contaminants will find their way into soil, ground water, and air. Once a hazardous material has leaked or been discharged to the surface, its movement depends on the nature of the material and the soil horizon. This chapter will discuss various means to sample and analyze both soil and ground water. Basic air quality sampling is covered in Chapter 3.

The viscosity of the contaminants and the permeability of the soil will determine how quickly the contaminants will move through the soil. A material such as gasoline with a low viscosity will be transported readily through sand which has a high permeability. However, the same contaminant gasoline will travel much more slowly through a clay formation which has a low permeability. In turn, a fluid such as used motor oil that is very viscous will be transported slowly through sand and more slowly through silts and clay. Gravity is generally the main cause of vertical movement of a contaminant through soil. This movement is controlled by the volume discharge, the depth to the water table and the density and viscosity of the contaminants. The vertical movement of a contaminant may often be stopped naturally if the contaminant encounters an impermeable soil layer or the contaminant is absorbed by the surrounding soils. Contaminants that are retained in the surface soils above the water table may exist as a liquid gas or be absorbed into the soil as a solid.

If enough contaminants are leaked or discharged onto the surface, under the right conditions they will eventually reach the ground water. Once the contaminant reaches the ground water its movements will be dependent on the physical properties of the contaminant. Two important properties to evaluate the contaminant flow in the ground water will be the specific gravity and the solubility of the contaminants. The solubility will help determine how fast the contaminant will be transported through the ground water while the specific gravity of the contaminant will help determine the extent the contaminant will move horizontally and/or vertically through the ground water.

For example, benzene, a gasoline constitute with a high solubility, will disperse readily in ground water compared to naphthalene which has a lower solubility. During investigation of gasoline and diesel spills, benzene, a gasoline constituent, will generally be found on the leading edge of the contaminant plume while naphthalene, a diesel constituent, is generally located closer to the source of the spill.

Contaminants which have a density greater than water and a low solubility will generally be found moving vertically through the aquifer and at higher concentrations at depth. In turn, a contaminant which has a density less than water and a high solubility will often be found in the upper section of the aquifer moving horizontally.

Once we understand the general principle of how contaminants are transported, we can evaluate the horizontal and vertical extent of the contaminated area. If the needed site specific data is not available, then a preliminary investigation is necessary. The preliminary investigation may require one or more of the following investigative methods: geophysical testing, soil boring and sampling, trenching/excavation and monitor well installation.

GEOPHYSICAL TESTING

Geophysical testing methods are of greatest benefit to locate and evaluate the horizontal perimeter of the contaminated area. Geophysical resistivity and conductivity meters are capable of evaluating subsurface layers, locating the water table and mapping contaminate contours. These surveys can be conducted easily and quickly by simply traversing the suspect contaminated area with the appropriate equipment.

Depending on site conditions, hand-operated conductivity and resistivity meters can evaluate site conditions from 20 to approximately 200 feet below land surface.

A magnetometer may be used to identify buried metallic objectives such as drums, tanks and other miscellaneous items.

SOIL BORING AND SAMPLING

Soil borings permit direct evaluation of soil samples collected from various depths. Soil samples can be examined visually for excessive contamination, or with field equipment during soil boring, or preserved for laboratory testing. Soil samples collected from the borings can also be evaluated for geologic conditions. The surficial soil above the water table should be characterized for horizontal soil layers which would affect the vertical and horizontal movement of contaminants through the soil to the water table.

TRENCHING AND EXCAVATION

Trenching and excavation permits direct observation of site-specific conditions. One or the other is generally preferred when solid wastes, of excessively contaminated soil, or free product will be encountered. Like soil borings, they allow for visual observation, field and lab analysis. The disadvantage of both excavation and trenching is that excavation depth is limited by equipment capability.

The simplest and most direct method to collect soil samples for evaluation is the use of a spade and scoop. A normal lawn or garden spade may be used to remove surficial soil to the required depth. After the surficial top soil is removed, a sample for field or laboratory analysis can be collected with a stainless steel scoop. However, this method is limited generally to sampling the first three feet of the soil horizon.

For depths greater than three feet, a post hole digger and hand auger can be used. The hand auger consists of an auger bit sampler, a series of drill rods and a T handle. The auger bit is used to bore a hole to the desired depth. This is accomplished by rotating the T bar and turning the auger vertically through the soil. Once the auger bit has reached the desired depth, the auger is retrieved and a soil sample collected. Soil samples can be evaluated in the field and submitted for laboratory analysis. Soil samples from depths below ten feet are often difficult and time consuming to collect with a hand auger. Therefore, soil samples below 10 feet are usually collected with a split spoon sampler attached to a drilling rig.

MONITOR WELLS

Once the contaminated area has been identified, monitor wells should be installed to evaluate ground-water conditions for dissolved contaminate and free product, permeability tests should be conducted and ground-water flow direction evaluated.

There are several common drilling methods available for well construction, including hollow stem augers, mud rotary, cable tools, air rotary.

Generally the hollow stem auger construction method is preferred for shallow monitor installing. The hollow stem auger drilling method is mobile and inexpensive to operate. If necessary, it is capable of drilling up to 150 feet into unconsolidated materials. Another advantage of the hollow stem auger drilling procedure is that it requires no drilling fluid. The addition of drilling fluid or other materials during the construction of the monitor well can alter laboratory analyses later.

Once the bore hole has been completed to the total depth, a small diameter well and casing, generally 2" by 4", is inserted inside the hollow stem auger. After the well screen casing has been placed in the bore hole to the total depth, the annulus is filled with sand or pea gravel. The sand or pea gravel acts as filter to remove fine sand and silt from entering the well bore. Above the sand or pea gravel filter, a bentonite or fine sand seal is installed to prevent cross contamination.

To properly evaluate the vertical extent of dissolved contaminants in the ground water, a well cluster may be required. A well cluster is generally two or more wells located near each other which penetrate different depths of the aquifer. Each well is screened at a different depth to obtain ground water samples at specific depths intervals.

For example the first well may be screened from 10 to 15 feet below surface, the second from 20 to 25 feet and the third from 30 to 35 feet. For most shallow water table aquifers, a 2–3 well cluster is appropriate.

To evaluate the horizontal extent of the dissolved contaminants in the ground water, a minimum of three wells is usually required. Generally one well is installed upgradient as a background well. The background wells are used to evaluate background contaminant levels, and to evaluate if contaminants from another source are transversing the contaminated area. A minimum of one well will be installed in the area of the highest suspected contamination and one well downgradient on the fringe of the contaminated plume.

Before a newly constructed well can be used for sampling, it needs to be developed. Well development is the procedure used to clear the well screen of fine silts and clays. The well should be developed until a sediment-free flow is achieved.

If sediment is left in the well, ground-water samples collected for laboratory analysis may not be representative of actual site conditions. Excessive levels of sediment in ground-water samples may alter future laboratory analysis.

After the contaminated area has been preliminarily investigated and monitor wells installed, ground-water samples are often collected. Ground-water samples are collected for laboratory analysis to confirm the type of contaminant and concentration of dissolved contaminants in the ground water. If free product is present, it is generally noted visually.

Before a well can be sampled, it must first be properly purged of standing water. Standing water is water that is not in contact with the flowing ground water. Standing water is often not representative of actual ground water conditions and also may interact with the well casing material and alter ground-water laboratory analysis.

Monitor wells are commonly purged by pumping a specific number of well bore volumes of water. While the well is purged, ground-water samples are collected and tested specific conductance, temperature and Ph. Once the ground-water specific conductance temperatures and Ph have stabilized, ground-water samples can be collected.

Ground-water samples can be collected with a wide variety of sampling devices. However, ground-water samples are generally collected with a bailer, suction pump or positive displacement pump.

The bailer is commonly constructed of teflon, PVC or stainless steel, with a check valve located at the base. The bailer is lowered into the well with a cable to collect a sample. When the bailer is lifted, the check valve is closed, allowing water in the bailer to be collected at the surface. Discrete depth bailers are designed with valves at both the top and base. This type of bailer is used to collect ground-water samples at discrete depths. The bailer is lowered into the well with a cable to the desired depth. Once the bailer is in position, the bailer valves are closed and a discrete sample is collected. Bailers provide an excellent and inexpensive means to collect ground-water samples.

The suction pump can be used to purge the well and collect ground water samples for laboratory analysis. As the name indicates, the suction pump operates by creating a vacuum in a sampling tube. The vacuum forces water in the well through the sampling tube to the surface. The suction pump has several disadvantages; one is that the pump is limited to 20 to 25 feet, depending on the pump. The suction pump may also cause aeration of volatile organics during sampling. Aeration of ground water samples may alter laboratory analysis.

A variety of positive displacement pumps are available to purge or sample monitor wells. The positive displacement pump is generally lowered into the well with a cable. The pump may be either pneumatic or electrical powered. The submerged positive displacement pump minimizes the aeration of ground-water samples. Therefore, ground-water samples collected for laboratory analysis are more representative of actual site conditions than samples collected with the suction pump.

Grab samples are generally the means of collecting liquid samples from trenches, excavations, ponds, small lakes or lagoons. Grab samples can be collected by attaching a precleaned laboratory container to a pole. Grab samples may need to be collected at various depths depending on site specific conditions.

The Kemmerer bottle is used to collect discrete grab samples at depths from surface water bodies or wells. The Kemmerer bottle is a messenger-activated water sampling device. In the open position, water flows easily through the sampling container. Once the container is lowered to the desired depth, a messenger is lowered down the sample line, tripping the release mechanism and closing the sample container. The Kemmerer bottle is an easy and inexpensive method of collecting discrete grab samples at depth.

LABORATORY ANALYSIS

To evaluate the extent and concentrations of contamination at a subject site, samples are often submitted for laboratory analysis. Because no single analytical method is applicable for all contaminants, past site operation and material/chemical storage on site should be investigated beforehand.

The preliminary investigation will often determine what analysis should be conducted. The most universal methods for trace contamination identification are the gas chromatography (GC), liquid chromatography (LC) and the mass spectrometry (MS).

Generally, contaminated samples to be evaluated with the GC can be volatile but can be gas, liquid or solid. High-molecular weight compounds generally can not be analyzed directly with a GC. The GC identifies contamination by retention time in a heated coil.

The MS analysis can evaluate samples much the same as the CS analysis. However, the MS analysis is used to confirm the structure of a compound. The MS analysis is a preferred method for analysis but is a more expensive test and requires an experienced operator.

The liquid chromatography is recommended for contaminants that are not recommended for GC chromatography. The high performance liquid chromatography can be used to identify contaminants that are less volatile.

To ensure that samples collected for laboratory analysis are representative of site condition, proper quality control procedures must be followed. Inaccurate laboratory data can lead to false conclusions concerning the type and degree of contamination at a site. In general, duplicate samples are required for proper quality control. Equipment blanks, field blanks, trip blanks and spike samples are collected for quality assurance.

The equipment blank is used to ensure that field sampling equipment has been properly decontaminated before sampling. The equipment blank will also help to ensure deionized distilled water used for decontamination has not been contaminated. The equipment blank is collected by pouring clean deionized/distilled water over or through the sampling equipment and collecting the sample.

The trip blank is used to evaluate container cleaning preparation, deionized/distilled (DI) water, sample cross-contamination during storage and transportation. Sample containers are filled by the laboratory with deionized analyte-free water. Trip blanks are usually picked up with empty sample containers and remain with the sample for analysis. Generally one trip blank for each parameter will be included with the sample containers.

The field blank is used to evaluate site conditions that would affect sample collection and analysis. The field blank may pick up airborne contaminants from exhaust fans or other industrial processes which are not representative of soil or water samples. Field blanks are collected by pouring deionized/distilled (DI) water into the precleaned laboratory containers. Field blanks should always be filled in the area where the sampling will be conducted.

Spike samples are samples with a known amount of contaminants added to DI water. Generally the spike sample is prepared by a laboratory or the sampling personnel, supplied by the laboratory. A known amount of contaminants is added to the spike sample to evaluate the laboratory's accuracy.

Duplicate samples are essentially identical samples collected from the same source under the same conditions and transported together. Duplicate samples are used to evaluate the accuracy of laboratory analysis. Generally one duplicate should be collected for every 10 samples collected.

Sampling and laboratory analysis are often the primary means to evaluate the health and environmental risk associated with a spill or discharge of hazardous or toxic waste.

Because improper sampling and analysis procedures can lead to misinterpretation of actual health and environmental risks, careful consideration and planning should always be given to sampling and analysis procedures to minimize inaccurate analyses.

ADDITIONAL REFERENCES:

Cahill, Lawrence, and Raymond Kane, *Environmental Audits, 6th Edition,* Government Institutes, Rockville, MD, 1989.

Chapter 6

• • •

AIR POLLUTION
CONTROL TECHNOLOGIES

Perry W. Fisher, Ph.D.
Kaushik Deb
Dames & Moore

OVERVIEW

Central to any environmental effort striving to reduce air pollutant emissions from industrial activity, air pollution control (APC) technologies play a very pivotal role. In 1991, projected APC system orders for North America alone totalled $3,860 million, for Europe $4,085 million, and for the rest of the world, $4,077 million.[1] Coupled with the passage into law of the most aggressive piece of U.S. environmental legislation to date—the Clean Air Act Amendments (CAAA) on November 15, 1990—this multi-billion dollar industry is on the brink of unprecedented growth. It is with evolving APC technologies that the goals of the CAAA regulations and rules will be achieved.

AIR POLLUTANT CHARACTERIZATION

Most industrial activities—from utilities, pulp and paper manufacture, steelmaking, automobile manufacture to small-scale businesses like chemical resin production, laundries and bakeries—produce airborne emissions of particulates or gases. Typically, particulates are classified as either SPM (suspended particulate matter), TSP (total suspended particulates) or simply PM (particulate matter). For human health purposes, the fraction of particulates which has been shown to contribute to respiratory diseases is termed PM_{10} (i.e., PM with sizes less than 10 microns). From a control standpoint, particulates can be fully characterized by their following attributes:

(i) Particle size distribution (typically determined by Bahco analysis).

(ii) Particle concentration or loading in the exhaust airstream (expressed as mg/m^3).

(iii) Some physical properties for specific control applications, e.g., particle density, resistivity, etc.

Gaseous pollutants are similarly characterized by chemical species identification, e.g., inorganic gases such as sulfur dioxide (SO_2), nitrogen oxides (NO_x), and carbon monoxide (CO) or organic gases such as chloroform ($CHCl_3$) and formaldehyde (HCHO). The rate of release or concentration in the exhaust airstream (in parts per million or comparable units) along with the type of gaseous pollutant greatly predetermines the applicable control technology.

Apart from the difference in types of air emissions, i.e., particulates and gases, it is important to note that under actual plant conditions both mixed-phase and phase transference of particulates and gases can occur. Thus, from a given source which needs to be controlled, air pollutant emissions often occur as both particulate matter and gases. In this regard, the phenomenon widely referred to as gas-to-particle conversion (GPC) occurs fairly commonly.

Under such circumstances, it is imperative that the proposed control should address only the phase of the pollutant of concern dictated by the requirements of the emissions reduction program. This is necessary because most control systems are designed to be pollutant-specific and fairly phase-specific. For instance, a system for particulate control will be designed to attain a specified efficiency for particulate matter reduction; in addition, it may or may not achieve some minimal reductions in gaseous emissions through a variety of physical and chemical processes. To convince a regulatory body that the same control is effecting an emissions reduction for gaseous pollutants will usually require appropriate source sampling data.

Besides phase, air pollutants are also characterized as:

(i)   Criteria Pollutants—PM, PM_{10}, SO_2, CO, NO_x volatile organic compounds (VOCs), and lead (Pb); and

(ii)   Hazardous Air Pollutants (HAPs)—includes, but is not limited to, many individual species of VOCs and PM.

Many criteria pollutants are fossil-fuel combustion derived, while HAPs can virtually be emitted from any segment of industrial activity.

IN-PROCESS AIR POLLUTION CONTROL

Generally, APC technology refers to end-of-pipe or add-on controls to reduce air emissions of specific pollutants. However, with the evolution of technology and a better understanding of pollutant formation mechanisms (e.g., combustion elements such as burner and furnace configuration and design, flame temperature, catalyst properties, chemical kinetics, etc.) it is presently possible for many facilities to retard the formation of air pollutants in place by means of oxidation and other chemical reactions, rather than add-on controls. The benefits of such an approach include:

- Elimination of the capital and O & M (operation and maintenance) costs of additional pieces of control equipment;

- Elimination of process perturbations due to transient shutdowns (for maintenance or otherwise) of add-on controls; and

- If the piece of control equipment has no recovery potential (solvents, cement dust, etc.), it is a production overhead in the plant manager's psyche.

However, the advocacy of in-process APC is severely hampered by the fact that the option is heavily dependent on commercially available and proven technology, and is also not readily amenable to retrofit situations. Further, the technology can be very process-specific and even plant-specific. Large corporations spend millions of research dollars fine-tuning their processes; inadvertently or deliberately this may result in abatement of specific pollutants. In many cases, details about these processes are considered to be proprietary information and protected under a confidentiality agreement with the state regulatory agency.

Various approaches by which an industrial facility may reduce its air emissions with in-process controls are described below:

A. *Combustion Reactions*: Most of the common fossil fuels fired for steam and/or power generation include grades of coal (bituminous, sub-bituminous, anthracite, lignite and others), natural gas and #2, #5, and #6 fuel oils. Other fuels may include bark, wood waste, sludge, pulp liquor and refuse derived fuel (RDF). This excludes incinerators which burn a wide variety of wastes and are mostly controlled through fuel preparation and post combustion emission control. In-process APC for combustion includes the following approaches depending upon specific target pollutants:

   (i)   Fuel Substitution—Based upon established and widely used air pollutant emission factors for different fuels[2], the descending order of so-called "clean" fuels for most criteria pollutants are—natural gas, #2 fuel oil, blends of other oils, and grades of coal. Thus, the most rational in-process control, if feasible, involves fuel substitution, use of dirtier topping fuels with a cleaner primary fuel, and other permutations so as to optimize fuel requirements, costs and air emissions.

   (ii) Combustion Modification—Historically, NO_x and CO have been very successfully controlled by combustion modification techniques. Although both pollutants are essentially combustion byproducts, it should be noted that NO_x and CO reduction schemes generally work against each other. Levels of CO which are formed as an intermediate product of combustion increase with most combustion-modification-based NO_x control strategies.[3] Tables 1 and 2 present prevalent combustion modification approaches and strategies for these pollutants for two different boiler and fuel combinations. Various combustion modification techniques used widely for reducing combustion-derived NO_x emissions are reduced NO_2 and O_2 levels, low excess air (LEA), reduced peak temperature, reduced exposure time, optimum burner design, and staged combustion.

Table 1

Combustion Modification Schemes for NO$_x$ and CO Control

Boiler: Pulverized Coal/Cyclone

Control Approach	Applicable Controls	NO$_x$ Control Feasibility	Commercial Availability
Decrease Primary Flame Zone Oxygen	• Low Excess Air (LEA)	Increase in CO	Available
	• Staged Combustion		N/A
	– Overfire Air Injection (OFA)	N/A for cyclone	
	– Reduced Air Flow to Burners	Increase in CO	Available
	– Burners of Service (BOOS)	Not effective	N/A
	• Low NO$_x$ Burners (LNB)	N/A for cyclone retrofit	N/A
Decrease Residence Time at High Temperature	• Flue Gas Recirculation (FGR)	Not effective	Available
	• Reburning	Effective in R&D not fully field tested	N/A
	• Burner Redesign	Not effective	Available
	• Load Reduction	Adverse operational impacts	Available

Note: Above unit is a standby and peak demand unit which experiences substantial load variations.

Boiler: Natural Gas/Watertube with Air Preheater

Control Approach	Applicable Controls	NO$_x$ Control Feasibility	Commercial Availability
Decrease Primary Flame Zone Oxygen	Low Excess Air (LEA)	Increase in CO	Available
	Staged Combustion	Not effective	Available
	Low NO$_x$ Burners (LNB)	Effective	Available
Decrease Residence Time at High Temperature	Reduced Air Preheat (RAP)	Significan loss in boiler efficiency	Available
	Flue Gas Recirculation (FGR)	Effective	Available
	Load Reduction	Not effective	N/A

Note: Above unit has modulated firing rates in response to variable plant steam demands.

Table 2

Post-Combustion Strategies for NO$_x$ and CO Control

Boiler: Pulverized Coal/Cyclone

Control Approach	Applicable Controls	NO$_x$ Control Feasibility	Commercial Availability
Post-Flame Region NO$_x$ Reduction	Selective Catalytic Reduction (SCR) – Ammonia	Excessive energy requirements for temperature elevation, performance degradation at variable loads.	Available
	Selective Non-Catalytic Reduction (SNCR) – Exxon Thermal DeNO$_x$ – NO$_x$ – out by Fuel Tech	Difficult to ensure residence time ~0.5 sec. at 1600 to 1900°F	Available

Note: Above unit is a standby and peak demand unit which experiences substantial load variations.

Boiler: Natural Gas/Watertube with Air Preheater

Control Approach	Applicable Controls	NO$_x$ Control Feasibility	Commercial Availability
Post-Flame Region NO$_x$ Reduction	(SCR) – Ammonia	Excessive energy requirements for temperature elevation, performance degradation at variable loads.	Available
	(SNCR) – Exxon Thermal DeNO$_x$ – NO$_x$ – out by Fuel Tech	Difficult to ensure residence time ~0.5 sec. at 1600 to 1900°F	Available

Note: Above unit has modulated firing rates in response to variable plant steam demands.

Also, continuous monitoring of the flue gases for either excess oxygen or CO, fuel analyses, fuel preparation, pressure and temperature measurements, flame appearance and even periodic stack tests will determine whether any combustion modification scheme is operating properly. For control of individual or total VOCs, temperature modulation (to prevent thermal degradation of the organics) and flares are often used for control purposes.

B. *Process Reactions: Various* process reactions which are employed for the purpose of in-process control include:

   (i)   Reactant substitution and/or reformulation;

   (ii) Better control of reaction environment, e.g., lower reaction temperatures;

   (iii) Use of passive catalysts which interfere with pollutant formation; and

   (iv) Solvent recovery and reuse, if possible.

With the understandable reluctance of most facilities to divulge any details pertaining to their processes, dissemination of information with respect to this approach is limited.

ADD-ON AIR POLLUTION CONTROL

A. *Particulate Control*: Historically, particulate control has been one of the primary concerns of regulatory agencies and industries alike, since emissions of particulates are readily perceived in short-range depositions of flyash and soot and in impairment of visibility. Differing ranges of control can be achieved with the following equipment:

   (i)   Cyclones—wet, dry, axial flow, multicyclones, etc.

   (ii) Fabric Filters—shaker type, reverse-air, pulse jet, etc.

   (iii) Wet Scrubbers—venturi type, packed bed, etc.

   (iv) Electrostatic Precipitators—field number types, hot-side, cold-side, etc.

Upon proper characterization of the particulate matter emitted by a specific process, the appropriate piece of equipment can be selected, sized, installed, and performance tested. It is beyond the scope of this chapter to detail specific types of equipment currently available; however, the fundamental principles of operation for each of the broad classes of particulate control devices are presented below:

(i) Cyclones—PM is removed by centrifugal forces generated by providing a path for the carrier gas to be subjected to a vortex-like spin. Cyclones are very effective in removing coarser fractions of PM. The equipment can be arranged in either parallel or series to both increase efficiency and decrease pressure drop. Figure 1 shows the schematic of a typical dust cyclone.

(ii) Fabric Filters—For industrial applications devices are typically designed with non-disposable filter bags. As the dusty airstream flows through the filter media (typically cotton, polypropylene, teflon or fiberglass), PM is collected on the bag surface as a dust cake. Fabric filters are generally classified based on the filter bag cleaning mechanism employed. Figure 2 shows a pulse-jet type fabric filter system.

(iii) Wet Scrubbers—A counter-current spray liquid is used to remove particles from an airstream. Device configurations include plate scrubbers, packed beds, orifice scrubbers, venturi scrubbers, and spray towers, individually or in various combinations. Wet scrubbers can achieve high collection efficiencies at the expense of prohibitive pressure drops.

(iv) Electrostatic Precipitators—ESPs operate on the principle of imparting an electric charge to particles in the incoming airstream, which are then collected on an oppositely charged plate across a high voltage field. The dust cake is then collected from the plate by striking it with rappers. The dust collection efficiency is a strong function of dust resistivity. Figure 3 shows a schematic of the principle of electrostatic precipitation.

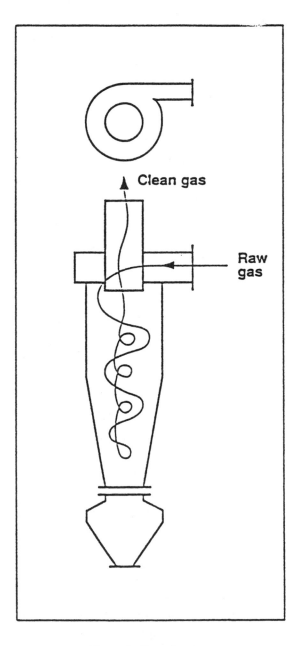

Dust Cyclone

FIGURE 1

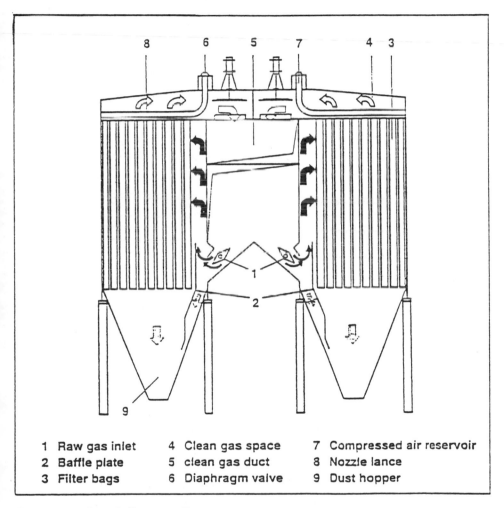

1 Raw gas inlet	4 Clean gas space	7 Compressed air reservoir
2 Baffle plate	5 clean gas duct	8 Nozzle lance
3 Filter bags	6 Diaphragm valve	9 Dust hopper

Courtesy: Lurgi Corporation

Pulse-jet Fabric Filter System

FIGURE 2

Tables 3 and 4 show typical characteristics, advantages and limitations of the above types of equipment. Sizing considerations, predictive equations for collection efficiency, failure rate, and economic considerations are important criteria which are discussed extensively in the literature.[8]

B. *Gaseous Pollutant Control:* Unlike particulate control a discussion of gaseous pollutant control does not lend itself readily to a description of equipment types, ranges of efficiency and pressure drop due to the dependence of gaseous pollutant control on the specific chemistry involved. Rather, a discussion of individual processes manifest in the various types of control equipment employed for these pollutants is more appropriate. In this regard, the four general processes used for gaseous pollutant control are:

(i)   Adsorption;

(ii) Absorption;

(iii) Catalytic Oxidation; and

(iv) Thermal Oxidation.

Each of the above processes is briefly described below:

(i)   Adsorption—This is a physico-chemical phenomenon in which the gas is concentrated on the surface of a solid or liquid. Subsequently, the captured gas can be desorbed with hot air or steam either for recovery or for thermal destruction. Usually, activated carbon is the adsorbing medium, which can be regenerated upon desorption. Adsorbers are widely used to preconcentrate a low gas concentration prior to incineration unless the gas concentration is very high in the inlet airstream. Adsorption also is employed to reduce odors from gases which have potential odor problems. The only major limitation for an adsorption system is the requirement for minimization of PM and/or condensation of liquids (e.g., water vapor) that could mask the adsorption surface and drastically reduce its efficiency. Figure 4 shows the schematic for an adsorption system.

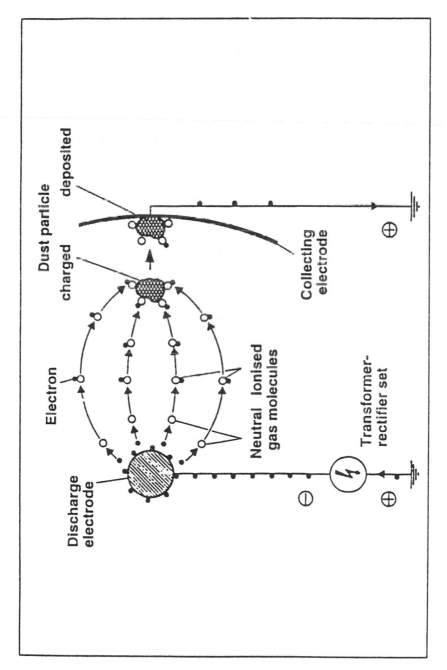

Courtesy : Lurgi Corporation

Principle Of Electrostatic Precipitation

FIGURE 3

Table 3

Typical Characteristics of Particulate Control Equipment

Particle Collection Device	Inlet Concentration (mg/m^3)	Typical Overall Collection Efficiency (weight %)	Pressure Loss (in. H$_2$O)	Maximum Gas Flow Rate (10^3 cfm)	Relative Space Required
Cyclone	2000	85	0.5–3	17.7	Medium
Multiple cyclone	2000	95	2–6	70.6	Small
Fabric filter	200	99	2–6	70.6	Large
Electrostatic precipitator	200	99	0.2–1	706.2	Large
Wet scrubbers					
— Gravity Spray	2000	70	1	35.3	Medium
— Centrifugal	2000	90	2–6	35.3	Medium
— Impingement	2000	95	2–8	35.3	Medium
— Packed bed	600	90	1–10	17.7	Medium
— Submerged orifice	200	90	2–6	17.7	Medium
— Venturi	200	99	10–30	35.3	Small

Source: ASHRAE, 1983.

Table 4

Advantages and Disadvantages of Particulate Control Equipment[9]

Advantages and Disadvantages of Cyclone Collectors

Advantages:
1. Low cost of construction.
2. Relatively simple equipment with few maintenance problems.
3. Relatively low operating pressure drops (for degree of particulate removal obtained) in the range of approximately 2- to 6-inch water column.
4. Temperature and pressure limitations imposed only by the materials of construction used.
5. Dry collection and disposal.
6. Relatively small space requirements.

Disadvantages:
1. Relatively low overall particulate collection efficiencies, especially on particulates below 10 μm in size.
2. Inability to handle tacky materials.

(continued on next page)

Table 4 *(continued)*

Advantages and Disadvantages of Fabric-Filter Systems

Advantages:
1. Extremely high collection efficiency on both coarse and fine (submicrometer) particulates.
2. Relatively insensitive to gas-stream fluctuation; efficiency and pressure drop relatively unaffected by large changes in inlet dust loadings for continuously cleaned filters.
3. Filter outlet air capable of being recirculated within the plant in many cases (for energy conservation).
4. Collected material recovered dry for subsequent processing or disposal.
5. No problems with liquid-waste disposal, water pollution, or liquid freezing.
6. Corrosion and rusting of components usually not problems.
7. No hazard of high voltage, simplifying maintenance and repair and permitting collection of flammable dusts.
8. Use of selected fibrous or granular filter aids (precoating), permitting the high-efficiency collection of submicrometer smokes and gaseous contaminants.
9. Filter collectors available in large number of configurations, resulting in a range of dimensions and inlet and outlet flange locations to suit installation requirements.
10. Relatively simple operation.

Disadvantages:
1. Temperatures much in excess of 288°C (550°F) requiring special refractory mineral or metallic fabrics that are still in the developmental stage and can be very expensive.
2. Certain dusts possibly requiring fabric treatments to reduce dust seeping or, in other cases, assist in the removal of the collected dust.
3. Concentrations of some dusts in the collector (~50 g/m^3) forming a possible fire or explosion hazard if a spark or flame is admitted by accident; possibility of fabrics burning if readily oxidizable dust is being collected.
4. Relatively high maintenance requirements (bag replacement, etc.).
5. Fabric life possibly shortened at elevated temperatures and in the presence of acid or alkaline particulate or gas constituents.
6. Hygroscopic materials, condensation of moisture, or tarry adhesive components possibly causing crusty caking or plugging of the fabric or requiring special additives.
7. Replacement of fabric possibly requiring respiratory protection for maintenance personnel.
8. Medium pressure-drop requirements, typically in the range 4- to 10-inch water column.

(continued on next page)

Table 4 *(continued)*

Advantages and Disadvantages of Wet Scrubbers

Advantages:
1. No secondary dust sources.
2. Relatively small space requirements.
3. Ability to collect gases as well as particulates (especially "sticky" ones).
4. Ability to handle high-temperature, high-humidity gas streams.
5. Capital cost low (if wastewater treatment system, not required).
6. For some processes, gas stream already at high pressures (so pressure-drop considerations may not be significant).
7. Ability to achieve high collection efficiencies on fine particulates, (however, at the expense of pressure drop).

Disadvantages:
1. Fossible creation of water-disposal problem.
2. Product collected wet.
3. Corrosion problems more severe than with dry systems.
4. Steam plume opacity and/or droplet entrainment possibly objectionable.
5. Pressure-drop and horsepower requirements possibly high.
6. Solids buildup at the wet-dry interface possibly a problem.
7. Relatively high maintenance costs.

Advantages and Disadvantages of Electrostatic Precipators

Advantages:
1. Extremely high particulate (coarse and fine) collection efficiencies attainable (at a relatively low expenditure of energy).
2. Dry collection and disposal.
3. Low pressure drop (typically less than 0.5–inch water column).
4. Designed for continuous operation with minimum maintenance requirements.
5. Relatively low operating costs.
6. Capable of operation under high pressure (to 150 lbf/in^2) or vacuum conditions.
7. Capable of operation at high temperatures [to 704°C (1300°F)].
8. Relatively large gas flow rates capable of effective handling.

Disadvantages:
1. High capital cost.
2. Very sensitive to fluctuations in gas-stream conditions (in particular, flows, temperatures, particulate and gas composition, and particulate loadings).
3. Certain particulates difficult to collect owing to extremely high- or low-resistivity characteristics.
4. Relatively large space requirements required for installation.
5. Explosion hazard when treating combustible gases and/or collecting combustible particulates.

(continued on next page)

Table 4 *(continued)*
6. Special precautions required to safeguard personnel from the high voltage.
7. Ozone produced by the negatively charged discharge electrode during gas ionization.
8. Relatively sophisticated maintenance personnel required.

(ii) Absorption—Absorption differs from adsorption in that it is not a physico-chemical surface phenomenon, but an approach in which the absorbed gas is ultimately distributed throughout the absorbent (liquid). The process depends only on physical solubility and may include chemical reactions in the liquid phase (chemisorption). Common absorbing media used are water, caustic, sodium carbonate, and nonvolatile hydrocarbon oils, depending on the type of gas to be absorbed. Usually, gas-liquid contactor designs which are employed are plate columns or packed beds.

(iii) Catalytic Oxidation—Predominantly used for the destruction of VOCs and CO, these systems operate in a temperature regime of 400 to 1100 °F in the presence of a catalyst. Without the catalyst the system would require much higher temperatures to operate. Typically, the catalysts used are a

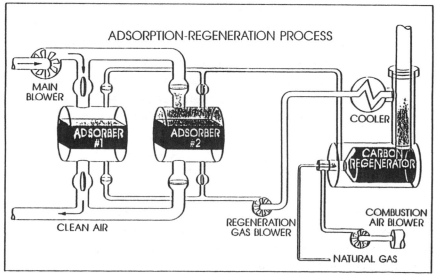

Courtesy : Calgon Carbon Corporation

Carbon Adsorption System

FIGURE 4

combination of noble metals deposited on a ceramic base in a variety of configurations (e.g., honeycomb-shaped) to enhance good surface contact. Catalytic systems are usually classified based on bed types such as fixed-bed (monolith or packed-bed) and fluid-bed. These systems generally have very high destruction efficiencies for most VOCs, resulting in the formation of CO_2, water, and varying amounts of HCl (from halogenated hydrocarbons). The presence of contaminants such as heavy metals, phosphorus, sulfur, chlorine, and most halogens in the incoming airstream act as "poison" to the system and can foul up the catalyst. Recently, however, a new generation of specialty catalysts (termed HDC and made by Allied Signal) containing a unique formulation has been developed which helps breakdown most hydrocarbons and halo hydrocarbons to CO_2 + H_2O + HCl in the operating temperature range of 700 to 1000 °F. Figure 5 shows a typical catalytic incineration system.

(iv) Thermal Oxidation—Without the use of catalysts most thermal systems operate at around 1500 °F or higher. Since the operating temperatures are roughly 500 to 1000 °F higher than catalytic systems, thermal units have a much higher auxiliary fuel requirement for preheating the waste gas stream. However, if the waste gas stream has a high calorific value (and thus participates in an exothermic reaction reducing heating requirements), the auxiliary fuel requirement is appreciably reduced. Generally, there is a trade-off between higher capital costs and catalyst replacement costs for catalytic systems and higher operating costs for thermal oxidation systems. Both systems, however, afford very high destruction efficiencies for VOCs and CO in optimal operation modes. Figure 6 shows a typical thermal incineration system.

Table 5 lists the advantages and limitations of the technologies discussed above. As with particulate control devices, design considerations, predictive equations for efficiency, and economic considerations are extensively discussed in the literature.[10]

Courtesy: Anguil Environmental Systems Inc.

Catalytic Incineration System

FIGURE 5

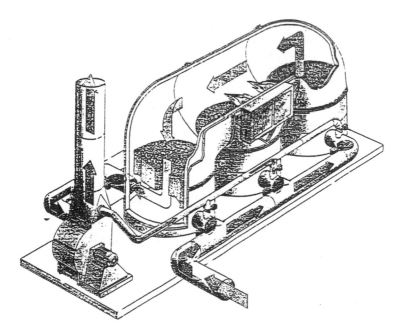

Courtesy: Salem Industries Inc.

Thermal Incineration System

FIGURE 6

Table 5

Advantages and Disadvantages of
Gaseous Pollutant Control Processes[11]

Advantages and Disadvantages of Adsorption Systems

Advantages:
1. Possibility of product recovery.
2. Excellent control and response to process change.
3. No chemical-disposal problem when pollutant (product) recovered and returned to process.
4. Capability of systems for fully automatic, unattended operation.
5. Capability to remove gaseous or vapor contaminants from process streams to extremely low levels.

Disadvantages:
1. Product recovery possibly requiring an exotic, expensive distillation (or extraction) scheme.
2. Adsorbent progressively deteriorating in capacity as the number of cycles increase.
3. Adsorbent regeneration requiring a steam or vacuum source.
4. Relatively high capital cost.
5. Prefiltering of gas stream possibly required to remove any particulate capable of plugging the adsorbent bed.
6. Cooling of gas stream possibly required to get to the usual range of operation [less than 49°C (120°F)].
7. Relatively high steam requirements to desorb high-molecular-weight hydrocarbons.

Advantages and Disadvantages of Absorption Systems
(Packed and Plate Columns)

Advantages:
1. Relatively low pressure drop.
2. Standardization in fiberglass-reinforced plastic (RFP) construction permitting operation in highly corrosive atmospheres.
3. Capable of achieving relatively high mass-transfer efficiencies.
4. Increasing the height and/or type of packing or number of plates capable of improving mass transfer without purchasing a new piece of equipment.
5. Relatively low capital cost.
6. Relatively small space requirements.
7. Ability to collect particulates as well as gases.

Disadvantages:
1. Possibility of creating water (or liquid) disposal problem.
2. Product collected wet.
3. Particulates deposition possibly causing plugging of the bed or plates.
4. When FRP construction is used, sensitive to temperature.
5. Relatively high maintenance costs.

(continued on next page)

Table 5 *(continued)*

Advantages and Disadvantages of Combustion Systems

Advantages:
1. Simplicity of operation.
2. Capability of steam generation or heat recovery in other forms.
3. Capability for virtually complete destruction of organic contaminants.

Disadvantages:
1. Relatively high operating costs (particularly associated with fuel requirements).
2. Potential for flashback and subsequent explosion hazard.
3. Catalyst poisoning (in the case of catalytic incineration).
4. Incomplete combustion possibly creating potentially worse pollution problems.

SUMMARY

Air pollutant emissions are primarily controlled either through in-process control or with add-on equipment. In-process control for combustion-derived pollutants is more widely used, and the various approaches on which it is based are either commercially available or readily accessible through information sharing. For process-related pollutants which can be very process- or even plant-specific, much less is known due to prevalent confidentiality clauses utilized by most facilities in submitting this information to regulatory agencies. With respect to add-on controls, a wide variety of such approaches exist for both particulate and gaseous pollutant emissions, most of which, if appropriately applied, can significantly reduce air contaminant emissions. Most add-on control equipment is widely commercially available.

ENDNOTES

1. R. W. McIlvaine, "The 1991 Global Air Pollution Control Industry," *J. Air Waste Mng. Assoc.*, v. 41, no. 3, March 1991, pp. 272–275.

2. *A Compilation of Air Pollutant Emission Factors: Stationary Sources*, USEPA AP-42, 1991.

3. *Sourcebook of NO$_x$ Control Technology Data*, USEPA, PB91-217364, July 1991.

4. R. A. Wadden and P.A. Scheff, "Engineering Design for the Control of Workplace Hazards," McGraw-Hill, 1987.

5. H. E. Hesketh, "Air Pollution Control—Traditional and Hazardous Pollutants," Technomic Publishing, 1991.

6. W. M. Vatavuk, "Estimating Costs of Air Pollution Control," Lewis Publishers, 1990.

7. Handbook of Control Technologies for Hazardous Air Pollutants, EPA/625/6-91/014, June 1991.

8. See notes 4-7 above.

9. *Perry's Chemical Engineers' Handbook*, 6th Edition, McGraw-Hill, 1984.

10. See notes 4-7 above

11. See note 9.

ADDITIONAL REFERENCES:

Brownell, William F. and Lee B. Zeugin, *Clean Air Handbook*, Government Institutes, Rockville, MD, 1991.

Control Technology for Hazardous Air Pollutants, Government Institutes, Rockville, MD, 1992.

Chapter 7

• • •

SOLID AND HAZARDOUS WASTE CONTROL TECHNOLOGIES

David C. Van Dyke
Dames & Moore

OVERVIEW

Throughout the existence of civilized man, the necessity of dealing with solid wastes has been a continual companion. As it is often found in the archaeology of long occupied sites, early civilizations simply buried themselves in the detritus of their culture. The past four thousand years have brought us sophisticated wastes and equally sophisticated methods and technologies for dealing with them.

The goal of this chapter is to present and familiarize the reader with various major technological fields which deal with solid and hazardous wastes. No attempt is made to give a comprehensive description of proven or potentially effective methods. To deal with solid and hazardous wastes is now a cradle-to-grave proposition. It is important that this be incorporated into the broader picture of industrial production and not be viewed as a separate entity.

Modern solid waste technologies have developed from the landfill. The landfill remains one of the most cost effective methods of disposing of solid wastes. The past 30 years have seen the landfill evolve from simply filling in the hole or gravel pit or developing lake (river) front to generate valuable new space for real estate. In the 1960s landfills began being sited in specific areas away from residential development. The technology of the day called for the landfill to be built on a sand base which would filter the water and leachate passing through it. As our awareness and knowledge of the leachate expanded, the performance of the landfill changed to its current "sealed capsule of waste" concept. Today's landfills not only have multiple medium liners such as clay and high density polyethylene (HDPE) but also require methods to remove the leachate generated and treat that liquid. The landfill also requires a battery of monitoring wells to assure that no leachate is escaping to the environment. The drawback to landfilling wastes is that they continue to exist. At some point in the future the landfill may leak or a regulatory requirement may be enacted which requires the owner (generator) to "reinvest" in the waste disposal process.

The technologies for treating solid and hazardous wastes have used the landfill as their springboard. One of the oldest technologies still in use is incineration. Initially, incineration was utilized as a volume and pathogen reduction technique, and this strategy remains a major factor in its utilization. The incinerator, as efficient as it is, became a target of environmental regulation early and today is one of the closest regulated technologies utilized. Incinerators, with their stacks, were visible and often detectable from their odor.

With the passage of the Clean Air Act the stack was capped, volatile emissions were scrubbed, and particulate was captured. With the development of Resource Conservation and Recovery Act (RCRA), the emissions and ash were often determined to be hazardous to the environment or population. The ensuing regulations have increased the cost of this technology, reduced the emissions, and limited the types of waste acceptable to this technology. Incinerators, like other successful technologies, have "spin-off" technologies that have adapted primarily to regulatory allowances. These include the utilization of the cement kiln and fuel blending as methods to reduce or eliminate a waste stream.

Another technology that has been in existence for a long time but has only recently gained widespread utilization is the stabilization of wastes. Stabilization by physical or chemical means prevents the waste from becoming mobile and causing impairment to the environment. Early stabilization techniques utilized the mixing of dissimilar waste products to physically tie up any potentially hazardous constituents. Examples of this exist in the ferrous foundry industry where melting emission dust was mixed with the metallic dusts from the shot blasting or the waste foundry sand containing bentonite clay. The resultant mixtures had performance results in leachability which allowed them to be disposed of through non-hazardous methods. Further stabilization methods included the addition of the waste into a new substance such as concrete, which when cured physically prevented contact between the waste and the environment.

Chemical stabilization accomplishes the same result as physical stabilization by creating an actual molecular change in the waste being stabilized. In this instance, the waste is changed into its naturally occurring state, which is non-leachable. An example of this is the metallic chlorides derived from melting process emissions, which when combined with sodium silicates will render the metallic silicate molecule which is non-leachable and therefore rendered non-hazardous.

Stabilization as a technology group is very effective in that the undesirable quality of the waste is the targeted area, i.e., liquid to solid, leachable to non-leachable. The end result is not waste reduction but is a waste that has manageable characteristics.

Another technology involves raw material recovery from solid and hazardous wastes. This concept was born from the banning of certain wastes from landfilling. In the case of emission dust from steel mill operations, the dust is passed through a high temperature plasma arc. In this process, the zinc is volatilized, captured, and recycled as a by-product. The remainder of the constituents are fused into a slag which is non-leachable and can be landfilled. The plasma arc process can be utilized on almost all solid wastes, but it is very expensive. This technology is another form of metamorphosis of a waste into constituents which may be recyclable or managed.

Solid and hazardous wastes are not always accessible; this is the case of contaminated soils. An emerging technology to deal with certain situations is the use of anaerobic microbes to perform a biochemical transformation of a contaminated area. The technology of "bugs" is rapidly expanding in both acceptance by the regulators and in the types of wastes that can be treated. The advantages of in situ treatment and inexpensive microbial populations are making biochemical treatment an increasingly popular and effective technology.

In some very select instances, solid hazardous wastes may be disposed of by "deep well injection" into the earth. In these instances the solid waste is mixed into a fluid form and pumped to a depth sometimes exceeding five thousand feet below the surface. This technique, although effective, is very expensive and receiving permit authorization to proceed will be more exhausting than most. The process of injection into the subsurface is really a "storage" of wastes rather than a disposal.

There also exists select situations where solid hazardous waste is allowed to be disposed by landspreading, that is, spreading the waste in a thin layer over the surface and later incorporating the waste into the existing soils. These last two possibilities are both site-specific and waste-specific but are brought out as possible alternatives. Physical, chemical, and biological breakdown of the waste occurs in the layer. Hydrocarbons are handled effectively in this manner.

When you are in the situation of dealing with solid hazardous wastes, remember that you will always have regulatory guidance and a plethora of regulatory enforcement personnel to assist you in the interpretation of those regulations. Other than the regulatory involvement, the technology to treat solid and hazardous waste varies little from that used for regular solid waste. Generally the controls are more stringent and the treatment is "tweaked" up a notch to assure completion of the treatment. Careful attention must be paid to the by-products of the process selected to insure that no toxic or other hazardous substance is created during the treatment of the original waste. Hazardous wastes are a fact of life that can be managed in an effective manner without anxiety. Key references for more information include documents from the following agencies:

U.S.E.P.A., Risk Reduction Engineering Laboratory, Cincinnati, Ohio

National Technical Information Service, Springfield, Virginia

Congressional Office of Technology Assessment, Washington, D.C.

ADDITIONAL REFERENCES:

Briese, Byron L., P.E., (ed), *Fire Protection Management for Hazardous Materials: An Industrial Guide,* Government Institutes, Rockville, MD, 1991.

Chapter 8

• • •

WASTE TREATMENT AND DISPOSAL TECHNOLOGIES

Robert G. Cooper
Dames & Moore

OVERVIEW

Waste disposal technologies have been developed in large measure as a response to the Resource Conservation and Recovery Act (RCRA) of 1976. This act, which is the primary legislation controlling hazardous waste management, initiated the concept of cradle-to-grave responsibility for waste management.

While RCRA is federal legislation, it permits states to develop their own hazardous waste programs as long as state programs meet, or exceed, the requirements of RCRA. The RCRA regulations set forth the protocol for determining whether waste is hazardous or non-hazardous, outlines the responsibility of the managers or handlers of the waste, and defines minimum standards of treatment for various waste groups.

The 1984 amendments to RCRA, implemented between 1984 and 1991, severely restrict land disposal prior to treatment of most waste streams. They tightened restrictions on land disposal over this period to promote more permanent and environmentally sound methods of waste treatment. They also authorized EPA to issue one-year permits for experimental facilities to assist with development of new treatment technologies.

Generators, transporters, storage facilities, treatment facilities, and disposal facilities are all regulated under RCRA.

The Comprehensive Environmental Response, Compensation and Liability Act (CERCLA), which is generally referred to as Superfund, provides government the authority to clean up contaminated sites. RCRA also contains provisions requiring the cleanup of sites where there is a potential hazard to the public. However, waste generated as a result of a Superfund cleanup must be managed in accordance with the requirements of RCRA.

HAZARDOUS WASTE TREATMENT AND DISPOSAL OPTIONS

Hazardous wastes are treated in tanks, biological reactors, surface impoundments, driers, incinerators, cement kilns, and boilers. These facilities may be either located and operated by a specific generator, or be the commercial type, accepting waste from a variety of generators.

Waste can be incinerated in a variety of incinerator designs, including rotary kiln, hearth, liquid injection, fume, fluidized bed, and infrared. The latter two varieties account for less than 5% of current capacity, while rotary kilns, liquid injection kilns, and industrial boilers and kilns have the potential to adequately treat a large percentage of the required treatment capacity; however, permitting, emission controls, and liability concerns have limited use of potentially available facilities.

Hazardous waste disposal facilities include landfills, land treatment units, surface impoundments, and injection wells. Of this class of facility, landfills have been limited to the type of waste that can be deposited without prior treatment as the 1984 RCRA amendments related to land bans have been implemented.

The last of these restrictions, which came into effect in 1991, has effectively shut down landfill disposal without prior treatment where treatment methods are available.

AVAILABLE TECHNOLOGIES

There are four broad classes of treatment technologies that are available—thermal, biological, physical, and chemical; some technologies combine two or more of these basic technologies.

The balance of this chapter will briefly describe a number of technologies available for the treatment of hazardous waste. The EPA has an Innovative Technology Program which tracks the development of emerging technologies that are in the developmental stage. However, the technologies described here are currently accepted and in general use for hazardous waste treatment.

BIOLOGICAL TREATMENT

Biological treatment of hazardous wastes is widely used in the chemical and other industries.

Two types of biological treatment processes are available for the treatment of waste:

- Aerobic processes (Treatment is in the presence of oxygen.)

- Anaerobic processes (Treatment occurs in the absence of oxygen.)

An example of an aerobic process is a conventional aeration. In this process, domestic wastewater is treated using activated sludge. Anaerobic treatment occurs in a simple septic tank.

Biological treatment is generally limited to organic wastes, although research is ongoing on biological treatment of some metals and metal compounds.

In aerobic treatment, organisms require both an energy and a carbon source for growth, and both affect what particular organisms will grow in a particular environment. Microorganisms will grow if provided the correct conditions, including carbon and an energy source. Many hazardous-waste streams satisfy either basic requirement and, if the appropriate nutrients are provided, a thriving organic population for waste treatment can exist. This will result in biological treatment if the temperature and pH are controlled, and substances which are toxic to the active organisms are eliminated. It is relatively easy to promote the required conditions for biological treatment in a fluid medium, but the same conditions are more difficult to achieve in contaminated soil and solids.

The most important aspect limiting the applicability of aerobic biological treatment of hazardous waste is the biodegradability of the waste. Biodegradability of a particular waste is very system-specific, and the correct conditions for successful treatment must be maintained to encourage the correct microbe mixture.

Aerobic treatment of fluid waste streams has been widely, and successfully, practiced for years in the chemical industry. Contaminated soils pose a more difficult problem, due to the difficulty of promoting homogeneous conditions. In-situ soil treatment has obvious advantages, but better control of the process can be achieved in a bioreactor. A bioreactor is a vessel where soils can be mixed and provide good controlled conditions. The economics of bioreactors in this application are limited, but in-situ treatment and land farming of some wastes has been successfully achieved.

ACTIVATED CARBON ABSORPTION

Activated carbon absorption can be used to treat a variety of contaminants in liquid or gaseous streams. This treatment method is most often applicable to organic compounds, with limited applicability to inorganic compounds.

Carbon-absorbing systems generally use granular activated carbon media in flow-through vessels which are designed for liquid or gaseous media. The absorption process is reversible and therefore it is possible to thermally regenerate the carbon for reuse which drives off the collected compounds. The waste compounds then can either be collected for reuse, or destroyed in the regeneration process. Nonthermal regenerative techniques are also used, including pH adjustment and steam.

There are three steps in the adsorption process. The molecule to be absorbed is transferred from the bulk gas of fluid to the exterior surface of the carbon. The molecule is then transferred to the interior of the carbon particle, through the pores, to an available site. The molecule is held to the surface in the final step.

Reactor designs vary from regenerative to disposable and have been tailored for almost any requirement. However, the adsorption process is not selective for specific organic compounds. Therefore, pretreatment to remove nonhazardous organics is desirable to prevent saturation of the carbon media with non hazardous compounds. This would limit the treatment capacity and effectiveness for target hazardous waste compounds. Suspended solids, oil and grease, and unstable compounds in the waste stream can all pose problems to a carbon treatment system.

ELECTROLYTIC RECOVERY TECHNIQUES

Electrolytic recovery is primarily used for the recovery of metals from process streams. Recovery techniques were developed for the mining industry but are currently used in electroplating and electronics industries to remove metals from wastewaters.

Electrolytic recovery is based on the oxidation-reduction reaction where electrode surfaces are used to collect the metals from the waste stream. The metal is collected at the cathode of the system, while gases such as hydrogen, nitrogen, and oxygen are given off at the anode. The gases produced depend upon the composition of the waste stream.

A typical recovery system consists of a treatment tank with electrodes, a power supply, and a gas handling (and treatment) system. Recovered metal must be removed from the electrodes periodically when the design thickness is achieved for the recovered metal.

This treatment technique is generally applied at a specific generator's site to recover valuable metals, to clean process waters, or to treat wastewaters prior to discharge.

AIR AND STEAM STRIPPING

Stripping is a process where dissolved molecules from a liquid are transferred to a flowing gaseous stream. In air stripping, the moving gas is usually ambient air, which is used to remove volatile dissolved organic compounds from liquids including ground and waste waters. The process is driven by the concentration gradient between the air and liquid phase equilibrium for particular molecules. Steam stripping uses steam as the gas phase, and the driving force is provided by the organic compound and the water equilibrium imbalance.

Both air and steam stripping have been widely used for the removal of hazardous organic compounds from liquid wastes. Air stripping is applied to more volatile organics, while steam stripping is effective over a wider range of less volatile compounds. Both technologies are 99% effective in treating applicable waste streams.

Air stripping is widely used in the treatment of petroleum and volatile organic solvent-contaminated groundwater, while steam stripping is widely used in industry for the treatment of generally higher waste concentrations of less volatile compounds, as well as solvent recovery.

Air stripping applicability is limited to dilute concentrations of aqueous waste streams with contaminant concentrations less than 100 mg/L. In order to prevent fouling problems in airstripping process equipment, pretreatment may be required. Suspended solids and any dissolved compounds prone to oxidation which will precipitate during the stripping process will cause fouling problems.

STABILIZATION AND SOLIDIFICATION

Solidification and stabilization are techniques used to reduce the mobility of hazardous compounds in waste to allow land disposal. The wider use of these technologies has become necessary as the 1984 RCRA amendments have been implemented which ban the disposal of a large percentage of hazardous waste without prior treatment. Where it is not possible to recover, remove, or convert hazardous compounds from a waste, the fixation of the hazardous compounds within a matrix allows landfill disposal.

Stabilization and solidification are treatment processes which improve handling or physical characteristics, result in a reduction of the solubility, or limit the leachability of hazardous compounds within a waste. Solidification, stabilization, encapsulation, fixation, and chemical fixation are all terms used to describe this class of technology.

Hazardous compound binding mechanisms are a critical factor to the success of this technology. Additives used as bonders include adsorption binders which reduce free phase liquid, pozzolan/fly ash and pozzolan/cement; thermoplastic binders physically trap compounds at the microscopic level. Macroencapsulation is used to encapsulate drums or blocs of waste in plastic or concrete containment.

In evaluating the applicability of solidification and stabilization technologies, the waste characteristic, the treatment objective, and the treatment evaluation requirements must be considered.

Treatment objectives fall into three categories. Level I involves removal of free liquids so that the waste can meet the physical test for landfill disposal. Level II treatment involves the removal of free moisture and evaluation for toxicity-characteristics leaching procedure (TCLP) which may result in the need to further treat the waste to reduce the mobility of hazardous compounds to acceptable levels. Level III treatment objective involves sufficient treatment to allow the waste to be delisted and classified as nonhazardous. The advantage of achieving Level III treatment is that the waste does not require disposal in a RCRA hazardous waste landfill.

THERMAL PROCESSES

Thermal treatment processes are used to treat both liquids and solids to either destroy the hazardous compounds or allow disposal of the process residue or treated waste in a RCRA hazardous waste landfill.

Thermal processes range from low temperature (800 °F) treatment of petroleum contaminated soils to remove volatile organics, to high temperature incineration and destruction of complex compounds such as dioxins at over 3,000 °F. Waste characteristics and treatment requirements determine the incinerator design to accommodate liquid or solid wastes. Temperature, turbulence, and retention time are the prime factors determining incinerator treatment design for both solid and liquid wastes.

Incinerator designs include more common types such as liquid injection and boilers, rotary kilns, fluidized beds and catalytics in both stationary and mobile configurations. More sophisticated and less common types of thermal treatment systems include wet oxidation, pyrolytic, plasma and molten glass processes.

FILTRATION AND SEPARATION

Filtration is the separation of solid particles from a liquid stream through the use of a semi-porous media. Filtration is driven by a pressure difference across the media caused by gravity, vacuum, centrifugal force or elevated pressure.

Filtration application to hazardous waste treatment falls into two categories—clarification of liquids where solids concentrations are less than 200 ppm, with the goal is to produce a cleaner effluent; and dewatering of slurries and sludges with the goal of concentrating the solids into a sufficiently solid form for further treatment or land disposal.

Filtration cannot be used to remove dissolved compounds unless these compounds are precipitated from solution through pretreatment. Nonmembrane filtration cannot separate compounds in the same phase, and sludges, tars, and other viscous compounds cannot be filtered effectively.

Filtration process designs begin with the simple granular bed gravity filter, and move toward more complex techniques, including pressure filters, vacuum filters, the plate and frame filter press, and the belt filter press.

Primary design considerations for filtration processes include waste characterization, viscosity, flow rate, temperature, feed variability, and treatment objectives.

The variety of filtration methods allow broad use of this technology to produce a clear effluent and to significantly reduce the moisture content of slurries and sludges to produce a dewatered product.

In hazardous waste treatment application, filtration may require the addition of pre- and post-treatment steps to achieve final waste treatment goals.

ULTRAVIOLET AND OZONATION TREATMENT SYSTEMS

This treatment technology is a developing technology with proven applicability for organic destruction in liquid waste streams. It is an advanced oxidation technology which uses a combination of ozone and ultraviolet light to treat hazardous organics in wastewater streams. Enhanced oxidation is a destruction technology where hazardous compounds are converted to nonhazardous constituents. This is a major advantage to the technology, as there are no residual hazardous residues after treatment.

Continuing development of this technology has overcome ultraviolet lamp fouling problems which reduced treatment effectiveness. The decomposition of aqueous ozone produces a hydroxide radical treatment with dissolved organic compounds in the wastewater stream. This technique has applicability for destruction of solvents, pesticides and other difficult to treat hazardous compounds.

HAZARDOUS WASTE LANDFILL AND SURFACE IMPOUNDMENTS

Hazardous waste landfills and surface impoundments are used for final disposal of treated hazardous waste which cannot be further effectively treated. These facilities contain and store hazardous wastes that are not currently recoverable or treatable.

These facilities are sited, planned, designed, constructed, operated, monitored, and maintained under strict federal, state and local regulation to safely contain hazardous waste residues.

Regulations fully control the wastes allowed to enter landfills and surface impoundments. These regulations restrict the land disposal of waste that can be effectively further treated prior to burial.

Landfills and impoundments are generally designed with a double synthesized liner system to contain the waste. The liner system is designed with leachate collection and leak monitoring systems to guard against the release of contaminants. Leachate generation is minimized in landfill cells through the restriction of semi-solid disposal, and a top layer system which eliminates surface infiltration into the waste cell. Landfill gas produced must be contained by the liner system, collected and treated prior to discharge to destroy hazardous compounds.

Landfills and impoundments are monitored through an approved complex perimeter and leak detection system which must be maintained after the facility is finally closed.

ADDITIONAL REFERENCES:

Hall, Ridgway M., Jr. et al., *RCRA Hazardous Wastes Handbook, 9th Edition*, Government Institutes, Rockville, MD, 1991.

Treatment Technologies, 2nd Edition, Government Institutes, Rockville, MD, 1991.

Chapter 9

• • •

SURFACE AND GROUND WATER POLLUTION CONTROL TECHNOLOGY

Neil J. Jost, Jr., P.E.
Dames & Moore

OVERVIEW

Like many technical endeavors, the practice of water and wastewater treatment has benefitted from numerous developments in science and engineering. Water and wastewater treatment has been, traditionally, more of an art than a science. With a history of development founded in experience, treatment is rapidly incorporating fundamental scientific and engineering principles. Advances in design and operation have progressed with better understanding of these fundamental principles.

Contemporary developments reflect the multiple efforts of the several scientific and engineering disciplines. Chemists, biologists, physicists and engineers in all disciplines have made substantial contributions over the last century. These contributions resulted in (1) identification and characterization of fundamental transport, reaction and surface phenomena (2) advances in materials science and (3) data and analyses contributing to policy decisions about environmental protection. This chapter presents, in brief, the technology spawned by this multidisciplinary effort as applicable to water pollution control.

Both surface water and ground water can be polluted. Surface waters receive discharges from a variety of sources: human wastes, industrial wastes, farms, mining and weather. Sometimes, spills of contaminants cause temporary adverse effects on surface waters. But rarely are these effects irreversible (consider the partial restoration of Prince William Sound after the *Exxon Valdez* oil release). Ground waters become contaminated from weeks or years of recurring spills or intentional above ground disposal, from leaking underground tanks and from intentional below ground disposal of harmful or hazardous materials in landfills. The cleanup of contaminated ground water, because it is below the ground surface and removed or modified slowly as part of soil or within an aquifer, may extend over weeks and years and decades. (See Table 1.)

Efforts to control water pollution in the last half of this century have paid off. Water quality in surface waters has generally improved as a result of government programs to control discharges since the early 1970s. Contamination of ground water in valuable aquifers received considerable notice in the early 1980s and is one of the outstanding contemporary problems for which considerable scientific and engineering effort is being mobilized. Pollution control technology has figured and will figure signifi cantly in the resolution of these problems: maintenance and protection of water quality in surface waters; and cleanup and restoration of quality of contaminated ground waters.

The challenge for cleanup of contaminated ground water is how to economically bring the problem and the solution together. First, geologists and hydrogeologists are seeking to better and more economically identify the extent and movement of contamination in the subsurface. Engineers face the problem of (1) getting treatment to the contamination and effectively treating pollution in place (in situ treatment) or (2) being able to retrieve and collect contaminated water and treat it at ground surface level. Defining the problem and framing the remedy are still elusive issues in view of the perceived lack of effectiveness of "pump and treat" remedies applied to date. "Pump and treat" refers to pumping ground waters to the surface where treatment then takes place.

TREATMENT TECHNOLOGY DEVELOPMENT

There are a variety of considerations in establishing appropriate and cost-effective technologies for pollution control. These include government-imposed standards, the chemical and physical nature of pollution, and the level of treatment cost effectively absorbed by the economy without major disruption of the economy.

Table 1

FACTORS AFFECTING QUALITY OF SURFACE AND GROUND WATERS

Contributor	*Factors Affecting Surface Water Quality*
Weather	Dissolved gases from the atmosphere Industrial air emissions including particles and gases Fugitive dusts from construction, mining, farming, etc. Pesticide residues Bacteria, mold, and other organic debris
Households	Human wastes Garbage and trash Detergents and oils
Industry	Oxygen demanding organic material Mineral residues Toxic and bioaccumulative organics and metals Acid and alkaline salts Oily waste
Agriculture	Fertilizer soluble residue Soil and minerals Pesticide residues Organic debris

Contributor	*Factors Affecting Ground Water Quality*
Weather	Soluble salts from earth, mineral, and surface debris Atmosphere and soil gases

Households	Septic tank discharges:
	    Human wastes
	    Mineral salts
	    Detergents
	    Trichloroethene septic tank cleaner
	Seepage from poorly conveyed surface wastewaters

Industry	Metals, salts, solvents, oils, and fuels
	    from dry pits
	    from wet wells
	    from underground tanks
	    from land disposal and surface dumping

Agriculture	Fertilizer salt residues Pesticide residues
Public Landfills	Solvents and petroleum derivatives Products of organic decay Metals and other inorganic ions

Standards imposed by government are both technology-based and risk-based. Discharges to surface waters must meet either a performance standard or a water quality standard. Ground water cleanup must often meet a drinking water supply standard.

Technology standards recognize the performance history of the selected technology and attempt to force consistent high level performance for classes of users. Water quality and drinking water standards are risk-based and designed to protect ecological and human health with ample margins of safety. They are set with only protection in mind and are relatively low compared to levels achievable with ordinary treatment schemes for surface discharges. As a result, the cost of meeting risk-based standards "drives" many technological developments. In time, pollution prevention built into industrial process development, in response to risk-based standards, may absorb the effort and resources currently spent on pollution control.

The nature of water pollution spans a broad range of physical and chemical parameters. Domestic wastes can be distinguished from industrial and mining wastes. Suspended solids can be distinguished from dissolved solids in water. Organic wastes can be distinguished from inorganic wastes. Readily degradable organic wastes can be distinguished from other organic wastes that are resistant and even toxic to the saprophytic microorganisms responsible for degradation in a treatment facility.

Oil-laden waters ("oil and water don't mix") are more difficult to treat than oil-free water. Radionuclides, although not common as water pollutants, present a unique situation for treatment. Tables 2 and 3 present the range of parameters and some key concepts toward understanding the applicability and limitations of treatment technologies.

To measure the effectiveness of a treatment technology, we use the term *cost-effectiveness*. Cost-effective is the term that can be applied to treatment technologies or systems and can be expressed as a cost per ton or pound of removal for the particular parameter. The cost components of treatment systems involve both capital and operating and maintenance (O & M) expenses similar to any industrial endeavor. Engineers typically select waste treatment alternatives for consideration and compare the life cycle costs for these systems. Different treatment technologies can be compared by their ability to remove constituents with standard economic evaluations used for capital costs and maintenance costs.

Table 2

Water Quality, Discharge and Ground Water Analytical Parameters

Water Quality Parameters of General Interest (supports a wide range of uses)

Coliform Bacteria (fecal coliforms)
Floating or Suspended Matter (Total Suspended Solids, Turbidity)
Color
Taste and Odor
Hardness
Alkalinity
Total Dissolved Solids
Iron and Manganese

Note: Water quality generally means those characteristics or range of characteristics that make water appealing and useful (useful also includes the meaning: nonharmful or nondisruptive to either ecology or the human condition within the broad spectrum of possible uses of the water). For example, a pleasing "feel" on contact or drinking and absence of odor, turbidity, or color are desirable immediate qualities. Imperceptible chemical characteristics also determine quality. The presence of toxic metals such as chromium or lead, excessive nitrogen and phosphorus (essential nutrients for plant growth) or dissolved organic material may not be readily perceived by the senses, but may exert substantial negative impacts on receiving waters and eventually human health. Impacts may range from a loss of esthetics to reduction in biological health reflected in loss of species diversity in the ecosystem to outright human health concerns.

Parameters of Interest Generally Applicable to Surface Discharges

Measurements of Organic Content	BOD5, Biochemical Oxygen Demand
	COD, Chemical Oxygen Demand
(all in mg/l)	Total Organic Carbon
Oil & Grease	Oil & Grease (mg/l, extraction/weight)
Solids$_i$	TSS, Total Suspended Solids, mg/l
pH$_i$	pH units
Nutrients	Nitrogen, all forms, mg/l Phosphorus, mg/l
Whole Effluent Toxicity	WET, bioassay, survival per cent

(continued on next page)

Table 2 *(continued)*

Parameters of Interest Applicable to Surface Discharges and Ground Water

Metals (and compounds)
    Chromium
    Antimony
    Beryllium
    Cadmium
    Copper
    Lead
    Mercury
    Nickel
    Selenium
    Silver
    Thallium
    Zinc

Inorganics and Anions
    Arsenic and compounds
    Asbestos
    Nitrate Nitrogen
    Cyanide

Organics
    Acenaphthene
    Acrolein
    Acrylonitrile
    Aldrin/Dieldrin
    Benzene
    Benzidine
    Carbon Tetrachloride
    Chlordane
    Chlorinated benzenes (other than dichlorobenzenes)
    Chlorinated Ethanes (including 1,2-dichloroiethane,
        1,1,1,-trichloroethane and hexachloroethane)
    Chloroalkyl ethers (chloromethyl, chloroethyl, and mixed ethers)
    Chlorinated naphthalene
    Chlorinated phenols (other than those listed elsewhere;
        includes trichlorophenols and chlorinated cresols)
    Chloroform
    2-chlorophenol
    DDT and metabolites
    Dichlorobenzenes (1,2-, 1,3-, and 1,4-dichlorobenzenes)
    Dichlorobenzidine
    Dichloroethylenes (1,1 and 1,2-dichloroethylene)

(continued on next page)

Table 2 *(continued)*

Parameters of Interest Applicable to Surface Discharges and Ground Water

Organics *(continued)*
2,4-dichlorophenol
Dichloropropane and dichloropropene
2,4-dimethylphenol
Dinitroltoluene
Diphenylhydrazine
Endosulfan and metabolites
Endrin and metabolites
Ethylbenzene
Fluoranthene
Haloethers
Halomethanes
Heptachlor and metabolites
Hexachlorobutadiene
Hexachlorocyclohexane
Hexachlorocyclopentadiene
Isophorone
Naphthalene
Nitrobenzene
Nitrophenols
Nitrosamines
Pentachlorophenol
Phenol
Phthalate esters
Polychlorinated Biphenyls
Polynuclear aromatic hydrocarbons (various)
2,3,7,8-tetrachlorobenzo-p-dioxin (TCDD)
Tetrachloroethylene
Toluene
Toxaphen
Trichloroethylene
Vinyl Chloride

Note: Taken from the priority pollutant list defined in the EPA and National Resource Defense Council Consent Decree, 1980.

Table 3

Chemistry Terms and Concepts

1. Concepts: "solubility," "in solution," "dissolved"

   A. Two or more substances that disperse themselves uniformly in all proportions when they are brought into contact are said to be completely soluble in one another, or completely miscible. The precise chemistry definition uses the words, "homogenous molecular dispersion of two or more substances".

   B. Examples
      i. All gases are completely miscible.
      ii. Water and alcohol are completely miscible
      iii. Water and mercury (in its neat silvery liquid form) are immiscible liquids.

   C. Between the two extremes of miscibility, there is a range of solubility; i.e. various substances mix with one another up to a certain proportion. In many environmental situations, a rather small amount of contaminant is soluble in water in contrast to the complete miscibility of water and alcohol. The amounts are measured in parts per million (a shot glass in a swimming pool).

2. Concepts: "suspension," "sediment," "particles," "solids"

   A. Often water carries solids or particles in suspension, e.g., finely divided sand in water. These dispersed particles are much larger than molecules and may be comprised of millions of molecules (the least common denominator of making a substance what it is). The particles may be suspended in flowing conditions and initially under quiescent conditions; but eventually gravity causes settling of the particles. The resultant accumulation by settling is often called sediment (in receiving waters) or sludge or residual solids in wastewater treatment vessels. Between this extreme of ready falling out by gravity and permanent dispersal as a solution at the molecular level, there are intermediate types of dispersion or suspension. Particles can be so finely milled (as in certain printing inks) or of such small intrinsic size as to remain in suspension almost indefinitely and in some respects behave similarly to solutions. Often, these intermediate dispersions present the greatest challenges to pollution control technology.

   B. "Emulsion"
      i. Emulsions represent a special case of a suspension. Oil and water do not mix. Oil and other hydrocarbons derived from petroleum generally float on water with negligible solubility in water. In many instances oils may be dispersed as fine oil droplets (an emulsion) in water and not readily separated by floating because of size and/

(continued on next page)

Table 3 *(continued)*

or the addition of dispersal promoting additives. Oil and, in particular, emulsions can prove detrimental to many treatment technologies and must be treated in the early steps of multi-step treatment process.

3.   Concept: "ion"

   A.   An electrically charged particle. For example, sodium chloride or table salt forms charged particles on dissolution in water: sodium is positively charged (a cation), and chloride is negatively charged (an anion). Many salts similarly form cations and anions on dissolution in water.

4.1.4 Quagliano, J. V., Chemistry(ital.), 2nd ed., Prentice-Hall, Englewood Cliffs, N.J., 1964.

TECHNOLOGY SUMMARIES

This section presents brief reviews of the various technologies that are available for treatment of discharges and decontamination of ground waters. Some of them achieve destruction of contaminants or eliminate undesirable or deleterious characteristics, others serve to remove and/or concentrate contaminants. The latter processes must always have an ultimate disposal repository (possibly reuse) for the removed or concentrated materials. The other processes may generate a residual waste which must be further treated and/or placed in an approved landfill. In general, the cost of handling and disposing of residuals is materially less than the cost of treatment of the primary liquid waste stream of concern; yet, in many cases, final disposal costs may figure significantly in the comparison of alternatives.

COAGULATION AND SEDIMENTATION

Gravity settling is one of the simpler pollution removal steps. Under still or dormant conditions, many kinds of solids will readily settle out of the water bulk of water. Settling is facilitated in large tanks (clarifiers) or ponds designed to accumulate solids for removal or natural breakdown. Settling usually precedes or accompanies other removal or destruction steps discussed below.

Fine particles and light solids often form stable suspensions. These suspensions can be destabilized to promote rapid settling under quiescent conditions using chemicals such as either mineral salts or synthetic organic polymers. This process is known as *coagulation*. Some suspensions of active biological solids have been described as producing surface chemicals like polymers which promote flocculation, growth and settling of particles that might otherwise remain in suspension.

For numerous industrial wastes, coagulation is preceded or utilized in conjunction with chemical reactions to remove dissolved metals or anions by *precipitation*. A reactant such as lime (calcium hydroxide) will react with dissolved aluminum or chromium to form insoluble particles or a "precipitate" identified as aluminum hydroxide or chromium hydroxide. Precipitation is often one of the more economic means to remove dissolved metals and other inorganics.

The residual solids produced are then thickened by holding under undisturbed conditions for a time. The thickened solids are usually dewatered by some mechanical means to draw off the water from the thickened solids. Some form of vacuum filtration or pressure filtration is employed to reduce the still significant volume of water tightly held to the solids before the solids can be landfilled or incinerated or destroyed some other way. Rarely is simple evaporation and drying of thickened solids practical.

A special but important case of interest is oil emulsion breaking. Acids, polymer, and heat treatment are often used singly or in combination to break emulsions. The agglomerated oil floats to the surface for removal, in contrast to settling. Some manufacturers have developed equipment that floats oils and solids in lieu of settling. This equipment is generically identified as dissolved air flotation (DAF). Treatment to remove oil often must precede other treatment processes to protect those processes from detrimental effects caused by oil. Oil can clog membranes, inhibit biological growth and coat equipment to the detriment of treatment performance.

BIOLOGICAL TREATMENT

Biological treatment utilizes nature's own readily available mixed culture of bacteria and other organisms to break down organic chemicals to water and carbon dioxide. Most often, these bacterial cultures are grown under highly oxygenated conditions with the addition of air via a mechanical aeration system. Less frequently used, but gaining in acceptance, is the cultivation of organisms without oxygenation. These anaerobic systems have as their advantage the production of methane gas and a high degree of solids destruction. Production of methane gas is valuable if it can be burned to replace ordinary fuels. Also, solids destruction reduces disposal costs.

A broad spectrum of natural and man-made organic chemicals can be destroyed by harnessing microorganisms. Acclimated cultures are necessary to consistently produce high quality water discharges. The maintenance of well-adapted bacterial cultures can be both a design and operational challenge. Acclimation is the result of *natural selection*, the proliferation and success of those organisms that advantageously use the supplied food (or energy) to reproduce and grow at the expense of others. Single organism cultures, rather than multiple, are beyond application to water pollution control at this time.

Other biological systems have been employed or studied. There have been reports of destruction of certain chlorinated organic compounds resistant to ordinary decay by utilizing fungus such as white rot fungus. Large waste treatment systems have incorporated natural systems using soil and plants to achieve complete assimilation of readily degradable and generally nontoxic organic wastes.

System designs can be briefly described using typical terminology. Suspended growth processes are referred to as activated sludge. Contact systems, where organisms grow on surfaces, usually are either trickling filters or rotating biological contactors. Oxidation ponds or treatment lagoons which combine both aerobic and anaerobic processes are also used. Land application systems may be characterized as "landfarming" and "cultivated wetlands." These systems can provide the necessary destruction in carefully managed programs at substantial capital and O&M savings.

Some of the systems described above, however, generate troublesome loads of solids to be dewatered and disposed or further destroyed. A variety of solutions are then employed to remedy the solids problem. Biological waste treatment sludges have been composted, landapplied and incinerated. In some cases, dewatered and well-stabilized solids are landfilled.

CHEMICAL OXIDATION

The action of undiluted household strength bleach spilled on dyed cotton is representative of the action of strong oxidizing chemicals on pollution. Within a short period of time, the bleach on dyed cotton removes or oxidizes all color at the point of contact, ruining the fabric. There are other strong oxidizers besides chlorine. These include both solid chemicals and gaseous chemicals like chlorine. Useful oxidizers include potassium permanganate, hydrogen peroxide, ozone, chlorine dioxide and sodium hypochlorite. Oxidizers are not readily substituted for one another as their strength and other direct and indirect effects guide their use.

Because of cost constraints that are due to long reaction times, none of the oxidation systems utilizing the oxidants listed above can be considered complete destruction systems. Some specific industry wastewaters and selected contaminated ground waters may be treated effectively because of the nature of the pollutants (in wastewaters) and the low level of contamination (in some ground water). Inorganic substances like manganese, iron, sulfur, cyanide and sulfite and organic substances like phenols, amines, humic acids, dyes, bacteria, algae and some toxic organic compounds can be "modified" effectively with oxidants to eliminate the objectionable or toxic character of the water without total and complete destruction of all substances.

A couple of examples may be helpful. Chlorine treatment for disinfection of water supply is one example of achieving useful effect (elimination of pathogenicity due to bacteria or other organisms) without complete destruction of organic matter. Color in industrial wastes due to dyes or pigments can be removed in some cases with ozone gas (ozonation). A limited amount of organic material is oxidized or destroyed; but key molecular bonds contributing color are broken. This same action may make the waste more amenable to biological treatment.

Design and operation requirements include the appropriate materials and vessels to hold the oxidants and wastes for reaction. Special materials of construction and safety features to prevent hazards to operating personnel are required and add to the cost of such systems. Specific situations and waste character may make chemical oxidation cost-effective. There is some production of solids with oxidation. Heavy solids require handling and disposal described above. Lighter solid loads may be removed by filtration (see discussion on polishing systems).

MEMBRANE PROCESSES

Membrane processes along with ion exchange exhibit more of a "high tech" quality than any of the other technologies described. The reason for this is the combination of developments in membrane and resin chemistry and the compact modular "black box" quality of the operating units. Membrane units have the potential to concentrate both fine particles in suspension and dissolved ionic or molecular species. Off-the-shelf equipment has the look and feel that often suggests one can turn it on and walk away from it. Unfortunately, membrane systems need careful attention to waste characteristics (less so for less contaminated water) so as to avoid plugging of the membranes and to achieve an effective level of concentration of pollutants.

Membranes are effectively groups of "gatekeepers" that function to allow only certain "acceptable" molecules through the gate. Pressure caused by special pumps is the driving force that forces the acceptable water molecules through the gate while the contaminants, either fine particles or dissolved species are retained. The chemical description of the process is concentration of the solute (the dissolved ions or molecules) in the solvent (water) using pressure as the driving force. The purified water is discharged or used. The concentrated stream may be further concentrated using evaporation and used, disposed, or possibly incinerated.

These systems must be carefully selected to avoid significant operating and maintenance costs. Membranes generally are organic polymers which have limited lifetimes and can be easily plugged. Natural organic substances or mineral salts are possible membranes but rarely find use. Examples of membrane use include desalination plants for producing drinking water, concentration of pigments in printing wastes, and supplementary water purification for industrial processes. Membrane processes include reverse osmosis, ultrafiltration and electrodialysis. The particle size, nature of the contaminant and membrane characteristics figure in process selection.

ION EXCHANGE

Ion exchange is probably most identified with water softening. There are numerous household installations which provide softened water. Softened water tends to foam easier and provides a "softer" less scummy or scaley feel than hard water. "Ion exchange" is rather descriptive of the process. Smaller non-troublesome or non-polluting ions such as sodium held by electrostatic forces within the exchange resin are displaced by larger, more objectionable ions such as calcium and magnesium in hard water. With time, all of the exchange sites are used up and the resin must be regenerated. Acids or bases corresponding to the positive or negative charge characteristic of the resin are used to regenerate the resin. Regenerant streams must undergo further treatment or concentration prior to disposal.

Ion exchange does not have general use for pollution control but has unique capabilities that can be applied to water supply and selected wastewater streams. Ion exchange technology often serves to recover or remove radionuclides where no other technology is adequate.

SORPTION PROCESSES

The term *sorption* is a general expression for both adsorption and absorption phenomena. These phenomena have been recognized now in most natural physical, biological and chemical processes. Even ion exchange described above represents a form of sorption. In general, sorption entails the movement of a component (ion, molecule, particle, etc.) from the water liquid phase to a second phase, usually a solid. Activated carbon is the most commonly encountered sorbent although there are others. Peat and other natural substances like certain clay soils exhibit sorption properties. Soils associated with ground water contamination may effectively attenuate organic or ion contamination through sorption within and on the soil particles.

The range of chemical pollutants differ in their tendency to be sorbed. Although organic compounds are usually the pollutants considered for activated carbon treatment, inorganic materials are also sorbed. The capacity of sorption of a given sorbent like activated carbon and the tendency of the substance to be sorbed are the essential engineering concerns for selecting activated carbon or other sorbent as a cost-effective treatment technology. Many organic compounds are readily sorbed at favorable efficiency (weight percent of sorbent) which makes selection of activated carbon an ideal solution for low level organic contamination removal that occurs in ground water. In addition, activated carbon has been used to remove various chlorinated organic solvents and polychlorinated biphenyls. For wastewater streams with more concentrated pollutants, activated carbon may serve as a polishing step for removal of toxicity or color after other treatment like biological or chemical treatment.

Activated carbon has a limited lifetime. Studies can predict the lifetime of the sorbent; also, breakthrough can be monitored with testing for the substance of concern in the discharge. Activated carbon must be regenerated or replaced. For small installations that might be used to remove toxic organics from ground water, spent carbon is sent to an incinerator. Carbon manufacturers may regenerate the carbon for large installations. A few larger facilities may regenerate carbon on-site as part of the total treatment process. Regeneration may involve heating to volatilize and oxidize retained organics or steam stripping to the atmosphere.

VAPOR PHASE SYSTEMS

The term *vapor phase systems* is used to designate a group of technologies that generally take advantage of the physical property of volatilization or evaporation as a useful method to remove or concentrate pollutants. In many instances these systems require heat input to promote volatilization. Distillation, evaporation, steam stripping and air stripping are examples. Some systems volatilize the contaminant of concern, while others boil off water leaving the contaminant residue behind.

Solvents like perchloroethylene (used in drycleaning) can be effectively air-stripped. Government agencies decide if the airborne vapor presents an air pollution control issue or not. In many cases, the rate of solvent release is an insignificant air pollution source or health hazard.

Some solvent recovery systems utilize steam stripping with condensation to concentrate solvents for reuse. Distillation and fractionation allow rather precise cuts of solvents to be accumulated but are otherwise similar to steam stripping. Heat input processes such as distillation and steam stripping are not commonly applied to water pollution problems, but in certain industrial settings, they prove to be necessary and economic.

Evaporation of water is generally not economical for most pollution problems, also; however, where the material to be concentrated has some value for reuse or recovery such as certain valuable plating baths or certain food products, evaporators with energy-efficient designs prove economically advantageous.

POLISHING SYSTEMS

Polishing systems can include activated carbon, ion exchange or even chemical addition as described above; but this summary will address only filtration. For the most part, filtration is applied after most of the pollutant load is removed from wastewater or used routinely as water supply purification to remove low levels of solids. The mechanism of filtration is complex and ranges from a buildup of solids and scum as a cake at the surface of a diatomaceous earth precoat filter to deep penetration and retention within a mixed media filter utilizing sand and anthracite coal.

Polishing filter systems can remove fluffy solids and turbidity which may have escaped coagulation and settling. Often, treatment systems must meet stringent limits for solids in discharges. Filtration technologies are applied as a safeguard to meet the solids limit.

SUBSURFACE CONSIDERATIONS

Treatment of ground water is complicated by the soil matrix containing ground water. Some of the basic phenomena used to advantage for wastewater treatment also occur in the soil. For example, water movement is not rapid (in most cases) because of some of these phenomena, e.g., resistance to flow across a membrane without driving force because of small pore size, etc. As mentioned, soil can also attenuate contamination because of sorption but only with impracticably large volumes of soil.

The lack of natural bacteria at ground water depth precludes biodegration as an active mechanism of destruction. Spontaneous organic chemical degradation is so slow in rate as to be negligible. Consequently, contamination can persist and remain for a long period without active intervention. Environmental scientists and engineers have attempted to devise ways to accelerate the progress of cleanup in several different ways. Solvent and fuel contamination seem to predominate as ground water contaminants. A variety of techniques have been tried to remedy these contamination problems with varying success.

Bioventing or just *venting* is a promising treatment option. A vacuum can be applied to soil and ground water toward extracting volatile components in the air movement above ground water. The addition of air can also promote bacterial degradation.

A few firms have cultivated proprietary bacteria for the breakdown of a variety of organic compounds. These organisms and supporting nutrients have been injected into the contaminated zones with encouraging success.

The success of all these methods, as of yet, cannot be predicted with certainty even with pilot or bench tests. Experience is still the guiding beacon for application of these techniques. Bioremediation is the current term to describe the variety of methods to breakdown organics in-situ or in surface modules.

The effectiveness of standard pumping and treating has a full range from zero to one hundred percent. To summarize, subsurface conditions are sometimes unpredictable and complicate both the identification and mapping of contamination. These unknowns in ground water quality and quantity complicate the use of water pollution-control technologies which are more easily applied to surface water.

ADDITIONAL REFERENCES:

Joint Committee, Water Pollution Control Federation and American Society of Civil Engineers, *Wastewater Treatment Plant Design*, Manual of Practice No. 8, WPCF, Washington, D.C., 1977.

Wastewater Biology: The Microlife, Water Pollution Control Federation, Alexandria, VA, 1990.

Weber, Jr., W.J., *Physicochemical Processes for Water Quality Control*, Wiley-Interscience, New York, 1972.

McKinney, R.E., *Microbiology for Sanitary Engineers*, McGraw-Hill Book Company, Inc., New York, 1962.

McGauhey, P.H., *Engineering Management of Water Quality*, McGraw-Hill Book Company, Inc., New York, 1968.

Madsen, E.L., "Determining In Situ Biodegradation," EST, 1662, 1991.

U.S. Environmental Protection Agency, *Treatment Technologies*, Government Institutes, Rockville, MD, 1990.

Ground Water Handbook, 2nd Edition, Government Institutes, Rockville, MD, 1992.

Arbuckle, J. Gordon and Russell V. Randle, *Clean Water Handbook*, Government Institutes, Rockville, MD, 1990.

Storm Water: Guidance Manual for the Preparation of NPDES Permit Applications for Storm Water Discharges Associated with Industrial Activity, Government Institutes, Rockville, MD, 1991.

Chapter 10

• • •

UNDERGROUND AND ABOVEGROUND STORAGE TANK TECHNOLOGIES

Andrew P. Schechter,P.E.
Dames & Moore

OVERVIEW

Presently, there are several million underground storage tanks (UST) and above ground storage tanks (AST) in the United States. These tanks are used primarily for the storage of petroleum or hazardous chemicals. Leaking storage tanks are a threat to the health and safety of the general public and the environment. The United States Environmental Protection Agency (USEPA) has previously estimated that 25% of the underground storage tanks and/or associated piping may be leaking their product into the environment.

Approximately 50% of the population of the United States is dependent on ground water as a source of potable water. In addition, fumes from product discharged into the soil and ground water may migrate beneath structures and through utility lines and subsequently cause fires or explosions. Owners and operators of tank facilities are liable for clean up costs and damages caused by their tanks. The average cost for remediating a site containing petroleum contamination to the soil and ground water is on the order of $100,000 to $250,000 depending on the lateral and vertical extent of the contamination and the required clean up target levels. In some cases the potential clean up cost may exceed the value of the property.

Over the past several years, the USEPA and most state agencies have enacted legislation aimed at minimizing the potential for discharges, detecting leaks from tanks systems, and remediating sites where discharges have occurred.

111

UNDERGROUND AND
ABOVE GROUND TANK SYSTEMS

A storage tank is a vessel and its associated piping that contains a product. It may be constructed of steel or fiberglass. If more than 10% of the tank is below ground, it is, from a regulatory perspective, an underground storage tank. If less than 10% of the tank is below ground, it is referred to as an above ground storage tank.

Old steel tanks may have been installed in the ground without a protective coating to resist corrosion. Often, part of the tank becomes negatively charged while other parts become positively charged. When this occurs, moisture in the soil becomes an electric conduit by which current flows. When this occurs, the steel begins to deteriorate and eventually leaks occur. Today, most tanks are protected in a variety of ways to prevent corrosion. They may be coated with a corrosion resistant material and/or have cathodic protection. Corrosion protection may consist of a bituminous coating or a reinforced fiberglass coating.

Cathodic protection uses a sacrificial anode which is more electrically active than the tank itself. As a result, an electrical current will flow to the anode rather than the tank, preventing deterioration of the tank.

Impressed current protects underground storage tanks by inducing a current around the tank through a series of anodes. Since the current flowing from the anodes is greater than the natural current between the tank and soil, the flow of current through the tank skin is prevented and the tank itself is protected.

Above ground storage tanks are often protected through the use of coatings to resist rusting and corrosion. In addition, above ground storage tanks are usually placed in structures to prevent discharge to the soil in the event of a discharge. These structures are referred to as secondary containment structures.

In addition to steel tanks, many tanks and piping systems today are constructed of fiberglass reinforced plastic. While fiberglass tanks are not susceptible to corrosion, they must be properly installed to prevent stresses which may rupture the tank and piping.

In addition to the tank and piping, most storage tanks systems contain overfill and spill protection to minimize the potential for accidental discharges during filling operations. Spills generally occur when a truck's hose is disconnected from the tank. While the spills may be relatively small, they may occur frequently and result in significant soil and ground water contamination. Overfill discharges occur when the hose continues to discharge into a full tank. This may result in discharges to the environment through pipe fittings and the vent pipe. Since filling occurs under pressure, these discharges can be large.

Spill protection is usually provided by a catch basin around the top of the fill pipe to contain the product and prevent its migration into the soil. Over spill protection is provided by automatic shutoff devices that shut off the flow of product when the tank is almost full. These devices are similar to those used at the pump of a gasoline station when filling an automobile car. When the tank is nearly full, the pump automatically shuts off preventing overfill and spillage onto the pavement or soil surfaces.

REGULATORY REQUIREMENTS

Regulatory requirements for underground storage tanks depend on whether the system is existing or a new installation. From a regulatory perspective, an existing installation is a system that was installed prior to 1988.

All existing UST systems must have overfill and spill protection. In addition, corrosion protection and leak detection systems will have to be installed in accordance with the following schedule obtained from the Federal regulations. Some states have requirements that are more stringent than the Federal regulations. Consequently, the appropriate state agency should be contacted to determine the regulatory requirements in a particular State. In any event, the compliance schedule insures that the oldest tanks, those with the greatest potential for failure, are addressed first.

All existing tanks must have corrosion protection and spill and overfill prevention devices installed by December, 1998. For those tanks that were installed prior to 1965 or the date of installation is unknown, leak detection devices were required to be installed by December 1989. By December 1990, those tanks that were installed between 1965 and 1969 were required to have leak detection devices. Tanks installed between 1970 and 1974 were required to have leak detection devices by December, 1991. By December, 1992 tanks that were installed between 1975 and 1979 are required to have leak detection devices. All existing tanks that were installed between 1980 and December, 1988 are required to have leak detection devices by December, 1993.

All piping (both pressured and suction) which was installed prior to December, 1988, is required to have corrosion protection by December, 1998.

In order to evaluate the integrity of the UST, existing tanks must be checked at least monthly. These monthly checks may be made by using automatic tank gauging devices, soil vapor monitoring systems, interstitial (secondary containment) monitoring systems, ground water monitoring wells or other monitoring devices.

If the tank has corrosion protection or an internal tank lining, monthly inventory control with tank tightness testing may be substituted for monthly detection testing for ten years subsequent to installing the corrosion protection or internal lining.

SPILL PREVENTION, CONTROL AND COUNTERMEASURES

In order to prevent the release of petroleum from tanks, a Spill Prevention, Control and Countermeasures (SPCC) program is required under Federal law for above ground storage tanks having a capacity greater than 1,320 gallons of oil or more than 660 gallons in a single above ground storage tank. While UST programs prevent the discharge of petroleum products into the ground water, SPCC programs prevent the discharge of petroleum products into the navigable water of the United States or adjoining shorelines. In addition, SPCC programs are designed to protect surface waters. SPCC programs regulate facilities with relatively large underground storage capacities while UST regulations cover the smaller capacity UST's at facilities such as gasoline service stations.

SPCC plans require the use of an impervious secondary containment to mitigate potential releases. The secondary containment may consist of an impervious concrete tank having a volume greater than the largest tank contained within unless it is enclosed in a concrete vault or a double wall tank. Consideration must also be given to protect the concrete tank from the accumulation of rainfall. If the structure does not contain a roof, it must be equipped with a manually controlled pump or siphon or gravity drain pipe and manual valve to remove storm water. All penetrations must be impervious to the product contained in the tanks.

TANK CLOSURE

In order to remove underground storage tanks, many regulatory agencies require conformance to a specific procedure to insure the safe removal, detection and if necessary remediation of petroleum discharges. Some agencies require field analysis of soil samples directly beneath the location of the former tanks, prior to receiving authorization to backfill the excavation. In addition, the installation of a monitoring well and ground water analysis may be required to properly perform a tank closure. Some regulatory agencies require prior notification and will have a representative on-site to observe the condition of the excavation upon removal of the underground storage tank. In some cases, the underground storage tank may be abandoned in place, provided it is cleaned and filled properly to prevent future contamination.

SOIL AND GROUND WATER REMEDIATION

In the event there is a release of petroleum products to the soil and ground water, there are a variety of effective technologies available to mitigate the situation. Soil remediation may take place at the location of the discharge. This is often referred to as in-situ remediation. An example of in-situ remediation called vacuum extraction is a technology commonly used at gasoline service stations.

As most of us are aware, gasoline is highly volatile and rapidly evaporates when exposed to air. Vacuum extraction is a process by which air is induced into the soil at the perimeter of a spill site and extracted from the center. This results in a flow of air through the contaminated soil. As the air circulates, the gasoline is volatilized and mixes with the air and is removed. Often, the gas/air mixture is passed through a carbon filter to remove the gasoline vapors and prevent their discharge to the atmosphere.

Diesel fuels are not as volatile as gasoline. As a result, vacuum extraction is not as effective. In-situ remediation of diesel fuel spills may include bioremediation. Bioremediation consists of chemically altering the contaminant into harmless carbon dioxide and water. Bioremediation usually consists of adding nutrients and air to either native bacteria or specially grown bacteria that causes them to rapidly multiply and feed on the contaminant and convert it to harmless gases. When the food source, in this case the diesel fuel, is exhausted, the bacteria population is unable to sustain itself and returns to normal size.

Ground water may be remediated by a process called pump and treat. Ground water wells are installed within the contaminated ground water plume. Pumps are installed within the wells. The contaminated ground water is pumped to an air stripper, which is essentially a tower with a fan on the bottom. The contaminated water is sprayed from the top of the tower, through an air column flowing upward from the fan. As the air meets the contaminated water, it volatilizes the gasoline from the water and discharges it to the atmosphere or to carbon filters for further treatment prior to discharge. The remaining ground water may contain low levels of gasoline. Treatment through carbon filters prior to discharge into a sanitary sewer system, will remove the remaining organic contaminants.

Another procedure to remove petroleum products discharged from leaking storage tanks into the soil is called thermal destruction. Thermal destruction requires high temperatures to volatilize the contaminants and convert them to harmless gases. Often, contaminated soil is passed through a cement kiln to permit the thermal destruction of contaminated soils.

ADDITIONAL REFERENCES:

Federal (EPA), State, and/or County Regulatory agencies administering a UST Program.

EPA, *The Interim Prohibition: Guidance For Design And Installation For Underground Storage Tanks.* EPA/530-SW-85-023. August, 1986.

EPA, *Straight Talk On Tanks: A Summary of Leak Detection Methods for Petroleum Underground Storage Tanks.* EPA/530-UST-90-012. August, 1990.

Rizzo, Joyce A., (ed.), *Aboveground Storage Tank Management: A Practical Guide,* Government Institutes, Rockville, MD, 1990.

Rizzo, Joyce A., (ed.), *Underground Storage Tank Management: A Practical Guide, 4th Edition,* Government Institutes, Rockville, MD, 1991.

Young, Albert D., (ed.), *Corrective Action Response Guide for Leaking Underground Storage Tanks,* Government Institutes, Rockville, MD, 1990.

American Petroleum Institute, Washington, D.C.

Chapter 11

• • •

POLLUTION PREVENTION
APPROACHES AND TECHNOLOGIES

Gary F. Vajda, P.E.
Dames & Moore

OVERVIEW

As we advance through the '90s, the development and implementation of an effective pollution prevention program will likely become the single most essential component of the successful corporate environmental program, with "success" being measured in terms of both compliance and costs. Furthermore, the most successful pollution prevention programs will be those that comprehend the entire manufacturing process, not just the wastes that are generated.

An Integrated Manufacturing and Environmental (IME) approach to pollution prevention offers the opportunity to achieve compliance and the dual benefit of operational improvement and a return on investment. It is accomplished by recognizing that waste is, in essence, a quality defect, and by adapting traditional manufacturing quality control analytical methods to also consider environmental factors. The techniques and procedures entailed by this procedure are applicable to both large and small waste generators, and allow the most effective use of various pollution prevention technologies and opportunities.

THE IME APPROACH

For many years, factors attributable to environmental issues were not considered significant components of productivity and operating costs; instead, management focused on labor, materials and rework. Today, however, environmentally driven costs ranging from waste disposal to future Superfund liabilities to the costs of expanding a production operation must also be considered by plant and corporate management.

An approach to accomplishing this is found in several of the techniques and methods used historically to analyze manufacturing operations for productivity and quality improvement. These methodologies include functional systems analysis; statistical sampling programs; design of experiments (DOE); traditional alternatives analysis; and structured decision-making techniques. When applied to the objective of pollution prevention, the IME approach allows the company to evaluate the entire manufacturing system, not just the components.

GETTING STARTED

Before beginning the technical aspects of the IME program, it is essential to set the stage for success. As with any potentially controversial program, good planning and organization can be the difference between achieving success or not.

The key to a successful waste minimization program is a company-wide commitment—from the management, the implementors, and from the employees. The first step in accomplishing this is to establish the overall goals of the program. Whether large or small, these goals must be acceptable, achievable and measurable. People involved must believe that where they are going is reachable, and that there is a verifiable means to document progress. Once set, these program goals can be incorporated into individual department and/or employee goals.

The second step is to identify one or more individuals to lead the team putting the program in place. This person must be familiar with the facility, with waste management and production technologies, and with product requirements. He or she must also have good rapport with both management and employees and be able to bring a diverse group of people together.

Finally, the project team must have input from many departments and specialties, including:

- Production;
- Facilities/maintenance;
- Process engineering;
- Quality control;
- Environmental;
- Research and Development;
- Safety/health;
- Marketing;
- Purchasing;
- Material control;
- Legal;
- Finance; and
- Information systems.

Often it is useful, or even necessary, to bring in outside assistance to provide expertise and/or a broader perspective. However, it is essential that key input and direction be provided internally.

TECHNICAL APPROACH

Technically, the implementation of an IME approach lends itself to a multi-phased program:

Phase I: *Baseline Characterization* defines the production process and ascertains the current regulatory situation. This effort involves the review and evaluation of environmental and production data, flow charts, layouts, throughput, cost data, etc. One of the key components of this characterization is the development of a material balance of the process or facility. An "as-is" model is developed to describe and integrate operational and environmental considerations.

Phase II: *Alternative Development* evaluates various alternative production improvement and environmental compliance options. Each alternative is modeled using the same techniques used in Phase I. Recommended alternatives based on structured decision-making techniques, associated risks, and costs are presented, and the selection of the preferred alternative is made.

Phases III (Design) and IV (Installation) are traditional engineering activities and therefore will not be expanded further in this discussion.

During Phases I and II, it is essential to realistically characterize the production operations and establish a credible baseline. This means that an accurate "material" balance for the facility or the process must be developed. This balance should include not only the raw materials and finished product, but also items such as wastes (of all kinds), rejects, labor, and production left-overs.

For simpler operations this should not be a problem, but surprisingly it can be. Regardless, the need for having a good handle on what is going in and coming out of the plant cannot be over-emphasized. A good starting point, when available, is the SARA Title III (Superfund Amendments and Reauthorization Act of 1986) Section 313 Form R. However, to this must be added the "non-material" aspects of balance, such as labor and equipment constraints.

It should be noted that the model described in Phase I only needs to be as detailed or as complicated as is required for a specific circumstance. A complex computer model is neither warranted nor appropriate for a simple plating operation, no matter how large. Conversely, to attempt to assimilate the necessary information for a large multi-operation manufacturing or chemical plant without the aid of at least a simple data software package is asking for headaches.

For the more complex operations, one useful analytical technique is Integrated Computer-Aided Manufacturing Definition Methodology (IDEF). This method, an example of "function analysis," permits defining any process by a series of functions or nodes. Each node can be described in terms of inputs, outputs, the controls or constituents on the function and the mechanisms of transformation.

The added value of a baseline cost model is to help identify those operations that are high in cost. Analytical resources can then be focused on those operations or variables that yield the greatest return, environmentally and/or operationally. Among the modeling tools that can be used is a software package called the Standard Assembly Line Manufacturing Industrial Simulation (SAMIS) which calculates a unit production cost for a process or system. The value can then be used to compare improvement alternatives developed in Phase II. Again, this is more appropriate for larger projects. Large or small, however, it is important to comprehend the "hidden" costs of permitting, regulatory-related labor, and future liabilities.

For smaller or more focused projects, individual processes and functions can be analyzed utilizing alternate analytical techniques. Proven methods include:

- Statistical sampling programs or charting of historical data;
- Taguchi and other Design of Experiment process diagnostic techniques; and,
- Traditional alternative analysis methods based on rate of return, unit cost, risk analysis or other decision criteria.

Once the baseline model is validated and viable alternatives are developed, structured decision-making plays an important role. Ideally, the favored approach to decision-making in an uncertain environment is a blend of qualitative and quantitative methods to maximize benefits. This can be graphically depicted for presentation purposes.

More specifically, an alternative's viability depends on its respective Environmental and Operational Improvement Potentials. These can be judged from a number of factors deemed appropriate by the user and may include such items as environmental and operational improvements.

Environmental Improvement:

- Amount and type of pollutant;
- Potential for disruption or legal action;
- Level of mitigation technology;
- Compliance safety margin provided by control technology; and
- Permitability of technology.

Operational Improvement:

- Contribution to increasing inventory turns or reducing lead times (speed);
- Reduction in set up times, product differentiation and mix (flexibility);
- Reduction of scrap and/or rework (quality); and
- Lower operating and capital costs.

With an "end of pipe" solution to an environmental compliance program, no benefit from operational improvement is typically achieved. In addition, negative cash flow is experienced from the purchase and operation of the new control technology equipment. However, a solution combining both operational improvement and environmental improvement can offer an increase in productivity with a return on invested capital.

OVERVIEW OF AVAILABLE ALTERNATIVES

For any given waste stream, there is a general hierarchy of management options, both technical and administrative. These include the following major categories:

1. Source Reduction;
2. Recovery/Re-use;
3. Waste Exchange; and
4. Treatment/destruction/disposal

Of these, only source reduction is generally considered to be true waste minimization, with recovery/reuse a preferred alternative.

Source reduction includes those techniques that result in an actual decrease in the quantity and/or toxicity of the waste generated by a given process. This should be the preferred management option where possible. The major methods of reduction commonly used include revising operating practices, product changes, and process modifications (changes in raw materials, technology, equipment).

A good example of **revising operating practices** to minimize hazardous waste is *source segregation*, which refers to the separation of listed hazardous wastes from nonhazardous wastes. Because of RCRA's "mixture rule," any waste that is a combination of listed hazardous and nonhazardous waste is classified in its entirety as a listed hazardous waste. In practice, this means that if a plant combines a small quantity of electroplating rinse water treatment sludge (a listed hazardous waste) with a much larger amount of nonhazardous biological sludge, the entire mixture would have to be managed according to RCRA requirements. This is true even if the mixture is itself characteristically non-hazardous. By keeping these two waste streams separate, the facility can significantly reduce the quantity of hazardous waste it produces. Other practices which should be reviewed include those affecting good housekeeping, inventory control, employee training, and spill/leak prevention and control.

Appropriate **process modifications** might include any of the following:

- Technology changes;
- Equipment changes;
- Improved control;
- Procedure changes; and
- Material substitutions.

Examples of this include the replacement of a solvent degreasing operation with an aqueous degreasing operation or even a dry abrasive system; the use of combustion rather than solvents to clean paint racks; and the addition of cooling coils to vapor degreasers to reduce solvent losses. Significant potential reductions can result from fundamental changes in processes, such as using dry powder paints instead of solvent-based paints. The primary consideration in evaluating a process modification is maintaining product quality.

Recovery/Reuse techniques are most appropriate for two general classes of wastes: organic waste, such as solvents, coolants, and waste oils; and metal-containing inorganic wastes. There is little overlapping of the processes applicable for each. There is a growing interest in the use of off-specification materials either as raw materials for similar processes or as feedstocks to the production of the same material. Potential technologies include:

- Distillation of solvents;
- Smelting of plating waste water treatment sludges;
- Concentration via evaporation or ion exchange of spent plating baths.

SUMMARY

For most companies, waste management is a reality that must be addressed in a logical and comprehensive manner. Pollution prevention will be a key part of any sound program. "End of pipe" solutions to environmental compliance problems offer little or no opportunity for process improvement and increased profitability of production processes. Through application of an IME approach, opportunities to capitalize on simultaneous improvement of both compliance and manufacturing operations are achieved. This is accomplished by classifying waste as a quality defect, and by applying traditional industrial engineering techniques used to improve productivity, quality and costs toward identifying appropriate technologies to solve environmental compliance problems.

ADDITIONAL REFERENCES:

United States Air Force, Wright-Patterson Air Force Base, *Integrated Computer-Aided Manufacturing (ICAM), Function Modeling Manual (IDEF). UM110231100,* (June, 1981).

Jet Propulsion Laboratory, Pasadena, California, *Standard Assembly-Line Manufacturing Industry Simulation (SAMIS) PC User's Guide, SAMIS Release 6.0,* (December, 1985).

Vajda, G.F. and Stouch, J.C., *"An Integrated Approach to Waste Minimization"* paper presented at Air & Waste Management Association Conference, Pittsburgh, Pennsylvania, (1990).

EPA 6002-88/025, *The EPA Manual for Waste Minimization Opportunity Assessments,* Hazardous Waste Engineering Research Laboratory Office of R&D, Technomic Publishing, 1990.

Paul Cheremisinoff, "Hazardous Waste Treatment and Recovery Systems, *Pollution Engineering,* (February, 1988).

USEPA 530 SW-87-026, *Waste Minimization,—Environmental Quality with Economic Benefits,* October, 1987.

USEPA 530-SW-89-049, *Waste Minimization in Metals Parts Cleaning,* 1989.

Paula A. Comella, "Waste Minimization/Pollution Prevention," *Pollution Engineering,* April, 1990.

Case Studies in Waste Minimization, Government Institutes, Rockville, MD, 1991.

Hazardous and Solid Waste Minimization and Recycling Report, Government Institutes, Rockville, MD, Yearly Subscription.

Facility Pollution Prevention Guide, Government Institutes, Rockville, MD, 1992.

Chapter 12

• • •

RECYCLING

Porter-C. Knowles, P.E. P.G.
Dames & Moore

OVERVIEW

As another one of these very fast-moving technologies, recycling technology is in a constant state of change. Not only are some of the technologies helping to make recycling more valuable, but political and economic pressures both help and hinder this entire process.

We have already addressed waste minimization (pollution prevention), which is a variation of recycling, and many of the same issues affecting that field also impact waste reduction efforts. There is, should be, or hopefully will be an economic incentive for recycling.

To begin with, it is important to state that *recycle is reuse*. For example, treatment facilities take formerly potable water from household use and treat it to a degree that it can be used for agricultural irrigation; that is recycling. Recycling doesn't require that the treated water again meet drinking water standards. Paper fiber that is treated and then reused as a mulch for seeds and fertilizer *is* recycling. The paper does not have to be recycled again into more paper to qualify as being recycled.

ECONOMIC INCENTIVES

One economic incentive to recycle has been the increasing cost of landfill disposal. The increased costs relate to tipping fees for depositing types of waste. Those fees reflect market demand, the actual cost of adding additional cells to a specific landfill, or the fact that there remains a limited, finite amount of space at the particular facility. At this point in the discussion, it should be noted that many landfills have become specialized. Some receive only particular types of waste such as organics/yard waste or special wastes, or hazardous waste. Obviously, the fees for special waste and hazardous waste are substantially greater than normal household wastes.

As you would expect, the tipping fees for typical waste in the urban east coast areas of the U.S. are probably among some of the highest in the country. Some of these fees may range from $70 to $120 per ton. In contrast, one might find midwestern states who have fees that may be on the order of $10 to $20 per ton.

With tipping fees in some cases, one order of magnitude higher in one location than another, it is not difficult to understand the saga of "trains of waste" sent from New York to the Midwest and Far West in an attempt to find a location for disposal. In fact, the lower tipping fees make this transportation of garbage by boat and by train an economic incentive for waste transportation. It should be noted, however, that proposed state and federal regulations are seeking to make tipping fees more universal across the country and to prevent these situations from occurring. Still, these events make headlines in our newspapers, magazines and on television on a regular basis.

COMPOSITION OF A TYPICAL LANDFILL

In general, commercial wastes comprise on the order of 60 percent of a landfill, with household wastes responsible for the remaining 40 percent. Of note and for discussion later in the chapter is the fact that about 60 percent of the commercial waste is probably recyclable. This fact is one reason why office recycling has been so successful. However, an additional high percentage eventually can also be taken from household waste.

In the May 1991 issue of National Geographic, there was a break-down of landfill space by volume in 1991. The break-down looks like this:

Breakdown of Landfill Space by Volume in 1991

1. Paper—50 percent—includes packaging, newspapers (as much as 18 percent), telephone books, glossy magazines and mail order catalogs (In 1970, paper took up 35 percent of landfill space)

2. Miscellaneous—20 percent—includes construction & demolition debris (15 percent), tires, textiles, rubber and disposable diapers (8 percent)

3. Organic—13 percent—includes wood yard waste and food scraps

4. Plastic—10 percent—includes milk jugs, soda bottles, food packaging, garbage bags and polystyrene foam (nearly 1 percent) (This has stayed about the same, thanks to thinner plastic)

5. Metal—6 percent—includes iron as well as aluminum and steel cans for food and beverages

6. Glass—1 percent—includes beverage bottles, food containers and cosmetic jars

MARKET INCENTIVE

In speaking to a major recycling firm in the U.S., the comment was made that their environmental manager spent over 50 percent of his time related to recycling in (1) trying to find markets for different recyclable materials and then (2) keeping those market buyers happy. Those companies that recycle materials in the current 1992 marketplace can afford to be very picky with respect to overall quality and type of material which they might use in their own individual processes of recycling or regeneration. Supply far exceeds demand and is a reason for low market prices.

It is almost intuitive that different materials will have a very different status related to market value in different locations around the country. If there have not been economic incentives to place a recycling facility in a particular area of the country, collection of that waste product may become a serious storage problem because of the cost to send it to another location for recycling. On January 17th of this year, the Wall Street Journal carried an article about the landfilling of a sizeable tonnage of glass because there was not a recycling facility that could handle that glass product except at much greater cost than in the landfilling process.

The fact is, a sizeable amount of capital expenditure is required to make recycling viable. The cost of collection, sorting and transportation is added to pretreatment and integration into the product stream. On a national basis, the supply side and collection of recycling materials has so far outstripped the ability of the marketplace to absorb those materials into a recycling or reuse process. Therefore the value of recycle feed stock material delivered to recycling facilities is at a price about one third less than what it was two to two and a half years ago.

While supply outstrips demand, many recycling companies demand and expect a much higher quality of sorted recycling feed stock than they used to. For example, glass bottles formerly used for pesticides and other toxic materials hopefully are screened by the recycling material collectors from businesses and homes. Some recycling bottle facilities will not take colored glass. Different grades of paper and boxes have also become an issue. One can imagine the contamination from pizza boxes with unused pizza inside. These additions won't be accepted by many facilities. Therefore, the collector of recyclables must go through several levels of effort to try and produce a quality and viable feed stock for sale. And even with great care, contamination of some plastic materials within a waste stream may make reuse of large batches of plastic unacceptable. Those batches, if failing tests at the recycling facility, must then be disposed of in a normal landfill.

RECYCLING AND REUSE MIND-SET

An important factor in the long-term viability of recycling is a specific mind-set. Different attitudes also prevail in different parts of the country. In the far northwest, the states of Oregon and Washington have long had extremely high returnable fees for recyclable materials. This attitude that recycling is extremely important and that the products resulting from recyclable material are not inferior to those of virgin material, is a critical factor in reuse/recycle long-term strategy. Since the 1970s, other areas of the country have also gradually been pushing the use of recycled material in products, and in some cases making recycled material a preference in bidding specifications for governmental acquisition. Some cities and local governments require their paper to be made of recyclable paper. It is this type of incentive that will continue to make further recycling and reuse economically viable.

RECYCLING WASTE TYPES

As earlier reported, paper constitutes about 50 percent of landfill space. Paper is roughly 75 percent recyclable but from an economic standpoint in the Midwest, economic benefits are marginal at best. These benefits have dropped because of the lowered price for recycled paper, the different standards today on types of paper used by recycling facilities (no color or slick inserts) and the costs of capital expenditures for preprocessing recyclables. For example, a "de-inking plant" is required for newsprint before it can become recycled paper.

Although the economic impetus may be variable and sometimes marginal, the fact is, recycling paper is certainly good for the environment. For every ton of material that is reused, one and one half to three tons of virgin materials are saved. Recycled paper can be produced from recycle fibers with 64 percent less energy and 58 percent less waste than from the virgin fibers. In addition, the production of recycled paper generates 74 percent less air pollution and 35 percent less water pollution than that of paper from virgin fibers.

Organic waste at about 13 percent of landfill space is also a key target for recycling. In many parts of the country, yard wastes are either limited or prohibited from local landfills. This trend of exclusion is gaining momentum. Since some estimates suggest that 75 percent of all existing landfills will be closed by the year 2000, easily recyclable yard wastes can be used as mulch and fertilizer. Many publications are available today on composting and on the use of mulch for landscaping and for minimizing erosion.

Plastics constitute about 10 percent of landfills, which is about the same percentage as in 1970. Plastic is imprinted with a triangular symbol and numbering from one to ten. This industry standard is now on most plastic disposables in the marketplace. Categories one and two are very recyclable, but recycling companies have been increasingly concerned with contamination. In many cases, contamination from pesticides, waste oil, and other toxic materials is absorbed in the plastic containers; and although they may be a number one or number two, these containers are not usable as feed stock. Higher numbered plastics may be recycled sometime in the future.

Metals are truly one of the better recycling success stories, primarily because of aluminum cans. Without question, the aluminum industry has not only supported but subsidized collection and reuse in this past decade. Although the present cost of reusable aluminum cans is down, this marketplace is viable and has been relatively consistent over the past few years.

"BUY RECYCLE"

In addition to the supply/demand factors, there need to be incentives to buy recycled goods. In many cases, this incentive can come from government procurement and preferential rules and regulations. This approach, however, may be punitive to some manufacturers. Certainly, industry association pressure to both manufacturers and users will be an important factor in making the recycling market viable. When all industries voluntarily push to meet certain guidelines without governmental prodding or restrictions, this approach is very effective. Users of packaging materials also provide an important incentive by their own procurement procedures.

McDonald's Waste Reduction Task Force, originated in 1990, was set up to outline waste reduction options for the company to consider. The outcome, however, was a 42-step action plan, a new environmental policy, and the widespread integration of waste reduction considerations throughout McDonald's operations. For example, chlorine-bleached paper products were switched to brown, unbleached paper or paper bleached by a more benign process wherever feasible. Brown paper bags made of 100 percent recycled materials (50 percent post-consumer) have replaced white bags. Oxygen-bleached coffee filters have replaced chlorine-bleached ones and the Big Mac packaging is now being made with unbleached paper. McDonald's currently is recycling corrugated boxes and asking suppliers to use corrugated boxes containing 35 percent recycled content, a new industry standard. Corrugated boxes make up over one third (by weight) of the trash generated at an average McDonald's restaurant. These facts are just part of an overall waste reduction and management goal system incorporated as ethic into company policy. These types of company standards will help make recycling material more economically viable.

Many other companies are also benefitting from waste prevention efforts. Herman Miller, Inc., located in Zeeland, Michigan, manufactures office and institutional furniture. The company has estimated that by utilizing reusable packaging and incorporating environmental protection at an employee level, the firm annually saves on the order of 1.4 million dollars.

OFFICE RECYCLING

Office recycling has been one of the most effective programs instituted in recent years to reduce landfill volume and increase recycling supplies. A number of federal and state and local programs encourage small and large businesses alike to develop an office recycling strategy as part of their regular operations. For example, the Greater Kansas City Chamber of Commerce is in preparation of a second edition of an office recycling pamphlet, a manual for small businesses. A substantial number of companies participate in the program, which includes recycling office paper, corrugated cardboard, newspapers, aluminum cans, plastic and glass.

THE BOTTOM LINE

Recycling in this country is at a critical juncture in large part caused by poor economic conditions across the country. This situation has hindered development of economic incentives to recycle greater volumes of material. One recycling company which got into the business in 1989 thought that the economics of recycling would turn around in three to five years. Today, they state that what they didn't know at the time was when that three to five years was going to start. It hasn't stated yet!

The fact is that recycled material collection has taken off across the country and received a great deal of support and encouragement. What needs to follow is a much greater industry and regulatory push to provide the incentives to utilize the feed stock supply source which is now and will be available in the future. Maybe lowered interest rates will help encourage capital expenditures for de-inking plants for paper recycling, as an example. Nevertheless, with environmental sensitivity on the rise and more and more segments of industry responding positively to recycling and reuse, this whole concept will soon become increasingly beneficial and profitable.

ADDITIONAL REFERENCES:

EPA has a variety of brochures and pamphlets relating to reuse and recycling. The EPA has a series of guides on pollution prevention for various industries such as commercial printing, pharmaceutical, etc.

Kansas City Chamber of Commerce, *Office Recycling, A Manual for Small Business*, 1992, second edition.

Complete Guide to Reducing and Recycling Your Office Waste, Government Institutes, Rockville, MD, 1992.

A number of recycling firms are listed in the Yellow Pages, many of which may have information related to this field.

INDEX

• • •

Also Available from Government Institutes

Clean Air Handbook
By F. William Brownell and Lee B. Zeugin
Softcover, 336 pp., 1991, $74 ISBN: 0-86587-239-2

Clean Water Handbook
By J. Gordon Arbuckle et al.
Softcover, 440 pp., 1990, $79 ISBN: 0-86587-210-4

Directory of Environmental Information Sources, 4th Edition
Edited by Government Institutes' Staff
Softcover, approx. 350 pp., 1992, $74 ISBN: 0-86587-326-7

Emergency Planning & Community Right-to-Know Handbook, 4th Edition
By J. Gordon Arbuckle et al.
Softcover, 192 pp., 1992, $67 ISBN: 0-86587-272-4

Environmental Audits, 6th Edition
By Lawrence B. Cahill
Softcover, 592 pp., 1989, $75 ISBN: 0-86587-776-9

Environmental Communication & Public Relations Handbook
By Bruce Harrison et al.
Softcover, 165 pp., 1988, $59 ISBN: 0-86587-748-3

Environmental Due Diligence Handbook, 2nd Edition
By William J. Denton et al.
Softcover, 300 pp., 1991, $74 ISBN: 0-86587-245-7

Environmental Evaluations for Real Estate Transactions: A Technical and Business Guide
Edited by Frank D. Goss
Softcover, 250 pp., 1989, $69 ISBN: 0-86587-765-3

Environmental, Health & Safety Manager's Handbook, 2nd Edition
Edited by Thomas F. P. Sullivan
Softcover, 242 pp., 1990, $59 ISBN: 0-86587-219-8

Environmental Law Handbook, 11th Edition
By J. Gordon Arbuckle et al.
Hardcover, 670 pp., 1991, $65 ISBN: 0-86587-250-3

Environmental Laws and Real Estate Handbook, 3rd Edition
By Steven A. Tasher et al.
Softcover, 290 pp., 1992, $74.50 ISBN: 0-86587-257-0

Environmental Management Review
Edited by Government Institutes' Staff
Quarterly, U.S. $188/Outside U.S. $220, ISSN: 1041-8172

Environmental Regulatory Glossary, 5th Edition
Edited by G. William Frick and Thomas F. P. Sullivan
Hardcover, 544 pp., 1990, $59 ISBN: 0-86587-798-X

Environmental Statutes, 1992 Edition
Hardcover, 1,165 pp., 1992, $57 ISBN: 0-86587-282-1

Environmental Telephone Directory, 1992-1993 Edition
Edited by Government Institutes' Staff
Softcover, 256 pp., 1991, $59 ISBN: 0-86587-278-3

The Greening of American Business: Making Bottom-Line Sense of Environmental Responsibility
Edited by Thomas F. P. Sullivan
Softcover, 350 pp., 1992, $24.95 ISBN: 0-86587-295-3

How the Environmental Legal and Regulatory System Works
By Aaron Gershonowitz
Softcover, 128 pp., 1991, $24.95 ISBN: 0-86587-244-9

Industrial and Federal Environmental Markets Report
Edited by Government Institutes' Staff
Three-ring binder, 320 pp., 1991, $185 ISBN: 0-86587-253-8

International Environmental Law Special Report
Edited by Government Institutes' Staff
Softcover, 400 pp., 1992, $85 ISBN: 0-86587-305-4

RCRA/Hazardous Wastes Handbook, 9th Edition
By Ridgway M. Hall, Jr. et al.
Softcover, 552 pp., 1991, $98 ISBN: 0-86587-270-8

Superfund Manual: Legal and Management Strategies, 4th Edition
By Ridgway M. Hall, Jr. et al.
Softcover, 442 pp., 1990, $95 ISBN: 0-86587-229-5

TSCA Handbook, 2nd Edition
By John D. Conner, Jr.
Softcover, 490 pp., 1989, $89 ISBN: 0-86587-791-2

Wetlands and Real Estate Development Handbook, 2nd Edition
By Robert E. Steinberg
Softcover, 218 pp., 1991, $72 ISBN: 0-86587-269-4

To order any of these publications or receive a complete catalog, please call: (301) 921-2355

or write:
Government Institutes, Inc.
4 Research Place, Suite 200
Rockville, MD 20850